AF724873

PDA/EMEA European Virus Safety Forum

Developments in Biologicals

Vol. 118

This series «Developments in Biologicals» begins with Vol. 102 and is the continuation of IABs symposia series «Progress in Immunobiological Standardization», Vols 1-5, «Immunobiological Standardization», Vols 1-22 and «Developments in Biological Standardization», Vols 23-101.

Organized and published by the International Association for Biologicals (IABs)

KARGER

Basel · Freiburg · Paris · London · New York ·
Bangalore · Bangkok · Singapore · Tokyo · Sydney

PDA/EMEA European Virus Safety Forum

Paul-Ehrlich-Institut, Langen, Germany
September 29 – October 1, 2003

Volume editor *James S. Robertson*
National Institute for Biological Standards and Control
(NIBSC), U.K.

Proceedings of a conference organized by the
PDA, and
The European Agency for the Evaluation of Medicinal Products
(EMEA)

30 figures and 33 tables, 2004

KARGER

Basel · Freiburg · Paris · London · New York ·
Bangalore · Bangkok · Singapore · Tokyo · Sydney

Developments in Biologicals

Bibliographic Indices. This publication is listed in bibliographic services, including Current Contents® and Index Medicus.

Drug dosage. The authors and the publisher have exerted every effort to ensure that drug selection and dosage set forth in this text are in accord with current recommendations and practice at the time of publication. However, in view of ongoing research, changes in government regulations, and the constant flow of information relating to drug therapy and drug reactions, the reader is urged to check the package insert for each drug for any change in indications and dosage and for added warnings and precautions. This is particularly important when the recommended agent is a new and/or infrequently employed drug.

Printed in Switzerland by Médecine et Hygiène, Genève.
ISBN 3-8055-7873-3

Programme Planning Committee

Patrick Celis, The European Agency for the Evaluation of Medicinal Products (EMEA), UK
Ernst-Guenter Graf, Pall GmbH, Germany
Albrecht Gröner, Aventis Behring GmbH, Germany
Paula Korhola, National Agency for Medicines, Finland
Johannes Löwer, Paul-Ehrlich-Institut, Germany
Gautam Maitra, PDA Europe
Philip Minor, National Institute for Biological Standards and Control (NIBSC), UK
James S. Robertson, National Institute for Biological Standards and Control (NIBSC), UK
Georg Roessling, Schering AG, Germany
Glenda Silvester, The European Agency for the Evaluation of Medicinal Products (EMEA), UK
Giuseppe Vicari, Italy
Hannelore Willkommen, Clearant, Inc., Germany

Organisation of the Meeting

Session I	Regulatory Requirements Chairs: *Jean-Hugues Trouvin*, AFSSAPS, France *Albert Farrugia*, Therapeutic Goods Administration, Australia
Session II	Testing Source Material Chairs: *Paula Korhola*, National Agency for Medicines, Finland *Giuseppe Vicari*, Italy
Session III (1)	Virus Validation, Virus Assay and Standardisation Chairs: *Hannelore Willkommen*, Clearant, Inc, Germany *Georg Rössling*, Schering, Inc., Germany
Session III (2)	Virus Validation, Model Viruses Chairs: *Hannelore Willkommen*, Clearant, Inc, Germany *Georg Rössling*, Schering, Inc., Germany
Session IV	Virus Inactivation/Removal Technologies: Application and Validation Chairs : *Johannes Löwer*, Paul-Ehrlich-Institut, Germany *Ernst-Guenter Graf*, Pall GmbH, Germany
Session V	Virus Safety Aspects of Advances Technology Medicinal Products Chairs: *James S. Robertson*, National Institute for Biological Standards and Control, (NIBSC), UK *Hannelore Willkommen*, Clearant, Inc., Germany

Introductory Remarks from the European Agency for the Evaluation of Medicinal Products (EMEA)

The manufacture and control of biological and biotechnological medicinal products has evolved from the early part of the last century. In the 1960s and 1970s, significant developments took place in manufacturing, purification and analytical techniques. In line with these developments, the regulation of these medicines also evolved, resulting in greater harmonisation amongst the EU Member States in the latter part of the 1980s and 1990s. This harmonisation was achieved through collaboration on the review of dossiers in the, then, new area of biotechnology, and the development of EU guidelines.

Virus safety of recombinant DNA, monoclonal antibody and plasma-derived medicinal products was one of the many topics discussed. Now at the beginning of a new century, it is appropriate to reflect on the experience gained and new issues to be addressed.

The PDA/EMEA European Virus Safety Forum was the first collaboration between the PDA and EMEA. The EMEA's objective for the Forum was to raise and discuss current issues related to virus safety, in order to provide a strong scientific base for further considerations by the EMEA's scientific committees, in particular CPMP and the Biotechnology Working Party (BWP). The Forum succeeded in meeting this objective thanks to the excellent speakers and lively participation from the attendees. The BWP work programme for 2004-2005, which can be found on the EMEA website (www.emea.eu.int), will take forward a number of issues discussed during the Forum.

John Purves
Head of Quality of Medicines Sector
Pre-authorisation Evaluation of Medicines for Human Use

Introduction

The viral safety of biologicals (e.g. plasma-derived medicinal products, recombinant products derived from cell cultures, monoclonal antibodies or products using new technologies for manufacture) is under continual review by regulatory agencies. Over the last 10 years, several guidance documents have been developed on a European and international level [1], setting standards for selection and testing of source materials as well as for investigating the capacity of manufacturing processes to remove or inactivate pathogens [2-4]. Both the safety of source materials and the capacity of processes to remove or inactivate pathogens are essential tools to assure virus safety of medicinal products for human use.

Based on the accumulated experience with the established regulations and the massive amount of data generated over many years, the PDA/EMEA European Virus Safety Forum was established to provide an opportunity to discuss current issues relating to the viral safety of recombinant proteins, monoclonal antibodies and plasma-derived medicinal products. The conference covered regulatory requirements for these products, viral testing of source materials, and several aspects of virus validation, such as the model virus approach as well as virus assays and their standardization. It provided an opportunity for regulators and industry members to air their views on unsolved problems, to highlight the need for more data to clarify the issues, and to assess the need for review and further development of current regulations.

New approaches such as "generic/matrix" validation of virus removal and inactivation were open for discussion. The examples presented at the conference illustrated a need to discuss these issues and to clarify the benefits and applicability of such approaches for specific product categories or technologies. Examples of "robustness testing" were provided and questions arose as to (i) whether general conclusions on the need for and extent of robustness testing for specific product steps can currently be drawn from the existing database and (ii) whether it is possible to give more specific guidance for robustness testing of virus removal or inactivation procedures.

Virus inactivation/removal technologies that are not broadly established in the industry were discussed and PDA's initiative to establish a system for the qualification of virus removal filters was presented. The final session focused on virus safety aspects of advanced technology medicinal products, including gene therapy, cell therapy, xenogeneic cell therapy and products from transgenic animals.

Forty speakers from Europe, the United States, Canada and Australia contributed to the success of this conference. More than 200 participants from all over the world enhanced the value of the conference by their participation in discussions and poster presentations. I wish to extend my heartfelt thanks to all the speakers and attendees, as well as the organizers who made this conference possible.

I am very pleased that this conference can be made available to a wider audience by means of this publication. It is my hope that this volume will contribute to general efforts to achieve and assure a high level of safety of medicinal products for human use, that it may inspire further discussion between industry and regulatory agencies in diverse areas, and finally that it will support the harmonization of regulations in this field.

References

1 International Conference for Harmonisation, including Europe, USA and Japan.

2 CPMP/BWP/268/95.

3 CPMP/BWP/269/95.

4 CPMP/ICH/295/95.

Hannelore Willkommen

Contents

Robertson J (ed): PDA/EMEA European Virus Safety Forum.
Dev Biol. Basel, Karger, 2004, vol 118, p 1.

SESSION I

Regulatory Requirements

Chairs: *A. Farrugia* (TGA)
J.H. Trouvin (AFSSAPS)

Robertson J (ed): PDA/EMEA European Virus Safety Forum.
Dev Biol. Basel, Karger, 2004, vol 118, pp 3-10.

European Regulatory Guidance on Virus Safety of Recombinant Proteins, Monoclonal Antibodies and Plasma Derived Medicinal Products

P. Celis, G. Silvester

EMEA, London, U.K.

Keywords: Virus safety, EU guidance, recombinant proteins, monoclonal antibodies, plasma-derived medicinal products.

Abstract: This article gives an overview of the principles of the European Regulatory guidance on virus safety and virus validation for products derived from mammalian cell lines (recombinant DNA products and monoclonal antibodies) and from human blood. It provides an insight of the current views of the European authorities on the virus safety aspects of these different kinds of products. For biotechnology (rDNA and mAb) products, the article elaborates on the experience gained within the centralised procedure and illustrates this by means of a survey of the virus validation studies of all biotechnology medicinal products authorised in the European Union from the beginning of 1995 to mid-2003. Trends in model viruses used in the virus validation studies and virus removal/inactivation steps for these products are identified in the survey.
For plasma derivatives, this overview illustrates how EU regulatory guidance on the virus safety of plasma-derived medicinal products is kept under review and links to the topics discussed during the PDA-EMEA Virus Safety Forum.

INTRODUCTION

The virus safety of medicinal products can be ensured by a combination of three parameters: (i) selection and thorough testing of the starting materials, (ii) evaluation of the inactivation and removal capacity of the production process and (iii) testing for freedom from detectable viruses during the appropriate steps of production. A combination of these three parameters is necessary to provide a sufficient level of assurance.

Two significant guidance documents have been issued, one at European level [1] and one at International (ICH) level [2]. These two guidance documents were developed in parallel and partially overlap: the ICH Q5A Note for Guidance covers all virus safety aspects of biotechnology products derived from human or animal cell lines, whereas the CPMP Note for Guidance has a wider scope, covering all biological products and providing guidance on the characterisation of effective steps for virus inactivation and removal as products derived from less well characterised sources (e.g. products derived from organs, tissues and blood of human or animal origin) are included.

Recombinant Proteins and Monoclonal Antibodies

Virus safety aspects

The three complementary principles to ensure virus safety (testing of source materials, evaluation of production process and virus testing during production) apply to products derived via recombinant DNA technologies (rDNA) and to monoclonal antibodies (mAb).

The testing of source materials for this category of products is two-fold: thorough testing of the cell lines used (master cell banks, working cell banks and end-of-production cells) and testing of all raw materials of animal or human origin used as cell culture components (e.g. bovine serum) or used during the purification steps (e.g. antibody capture resins).

Detailed guidance on the testing of the cell lines can be found in the ICH Q5A Note for Guidance [2]. Specific guidance exists for the testing of some raw materials of human or animal origin: reference is made to the Note for Guidance on the use of bovine serum in the manufacture of human biological medicinal products [3] and the Note for Guidance on plasma-derived medicinal products [4]. It is recommended that the use of animal or human raw materials be avoided, where possible, during the fermentation and purification of rDNA products and mAb. European authorities nevertheless acknowledge that in some cases it will be difficult or impossible to avoid the use of such raw materials. Therefore, if using a component of animal origin, it is recommended using a component that has been subject to virus inactivation or removal procedures. The use of products that have not undergone such inactivation or removal procedures carries an additional risk, which would be considered by the European authorities during the marketing authorisation procedure. Components of human origin (e.g. human albumin) require being subjected to virus inactivation and/or removal procedures. Although outside the context of virus safety, the EU guidance on transmissible spongiform encephalopathy agents (TSE) [5] and the EMEA position paper on CJD and vCJD [6] also need to be taken into consideration.

The virus inactivation and/or removal capacity of the manufacturing process has to be evaluated. Guidance on how to perform and interpret such virus validation studies, and the aspects to be taken into account when designing such studies, is given in the two Notes for Guidance listed in the Introduction [1,2].

The CPMP Note for Guidance [1], published in 1996, indicated that, where the starting material posed a low virological risk, such as a fully characterised cell bank, the purification process often did not contain a specific virus inactivation or removal step and a validated purification process was considered as providing sufficient levels of virus inactivation/removal. However, some manufacturers of mAbs were introducing specific virus inactivation or removal steps into their production process, since mAb-producing cell lines of murine origin secrete retroviruses. The ICH guidance [2] states that "If little reduction of infectivity is achieved by the production process, and the removal of virus is considered to be a major factor in the safety of the product, a specific, additional inactivation/removal step or steps should be introduced."

Both the ICH and CPMP guidelines indicate that partitioning steps, such as chromatography, can contribute to the removal of viruses, but it is difficult to control the number of variables and to scale down the process for virus validation studies. Furthermore, the partitioning might be dependent on specific physico-chemical properties of a virus.

Survey of virus validation studies of recombinant proteins and monoclonal antibodies authorised in Europe between 1995 and 2003

Introduction

Since January 1995, all products made using rDNA technology and monoclonal antibodies have to be submitted for licensing under the Centralised Procedure. This procedure is coordinated by the EMEA and as such the EMEA had the opportunity to conduct a survey on the virus safety evaluation of all biotechnology products authorised in Europe from January 1995 until July 2003. Information on products authorised via the Centralised Procedure is published on the EMEA Website (www.emea.eu.int) in the European Public Assessment Reports (EPAR).

Methods

The survey is based on the proprietary and confidential information held by the EMEA that formed the basis for licensing. The survey focused only on 'biotechnology medicinal products' i.e. medicinal products containing monoclonal antibodies or recombinant DNA proteins as active substances. Excluded from the survey were vaccines, blood products, medicinal products including chemical entities as active substances, and those products for which the marketing authorisation has been withdrawn.

It has to be kept in mind that the same active substance is sometimes included in different finished products. The formulation of these finished products can be different, resulting in different bio-availability profiles (e.g. fast versus long-acting human insulins); in other cases, the marketing authorisation applicant has filed a duplicate application, with the identical finished product (but a different invented name). In the survey, such duplicate or multiple applications are counted as a single product. The survey was intended to be exploratory only; therefore, no statistical analysis was performed in this survey.

Results

In total, 71 biotechnology medicinal products were authorised between January 1995 and July 2003. This corresponds to 56 different rDNA or mAb active substances (see Table 1). As this survey relates to virus safety aspects, the active substances manufactured in a bacterial or fungal host cell system (n = 23, 41%) will not be considered further; for these products there should normally not be any virus safety issues (provided no animal- or human-derived materials are used during the fermentation or purification processes). Further analyses were performed on the substances produced in animal or human cells lines or via the production of ascites (n = 33, 59%). For four of the 33 products (all four authorised before January 1997), the documentation that could be readily retrieved was not sufficient to include in the analysis ; hence they were excluded from the survey. Therefore, the analyses were performed on 29 products.

Table 1: Overview of the different host cell systems.

Host cell system	Number of authorised products	Number of product according to counting rules*
Escherichia coli	22	16
Saccharomyces cerevisiae	13	7
Chinese hamster ovary cells	23	21
Other mammalian cell line	7	6
Human cell line	4	4
Hybridoma/ascites production	2	2
Total	**71**	**56**

* This corresponds to the number of different active substances.

The inclusion of validated virus filtration steps and/or validated virus inactivation steps in the manufacturing process of rDNA and mAb was analysed (Fig. 1). In products authorised until July 2003, specific virus filtration and virus inactivation steps were included in 66% (n = 19) and 55% (n = 16), respectively, of the purification processes. For the years 2000 to 2003, the corresponding figures are 75% and 56% respectively. The most commonly used (validated) virus inactivation step is low pH treatment (8/29), followed by SD treatment (2/29) and Triton X-100 (2/29).

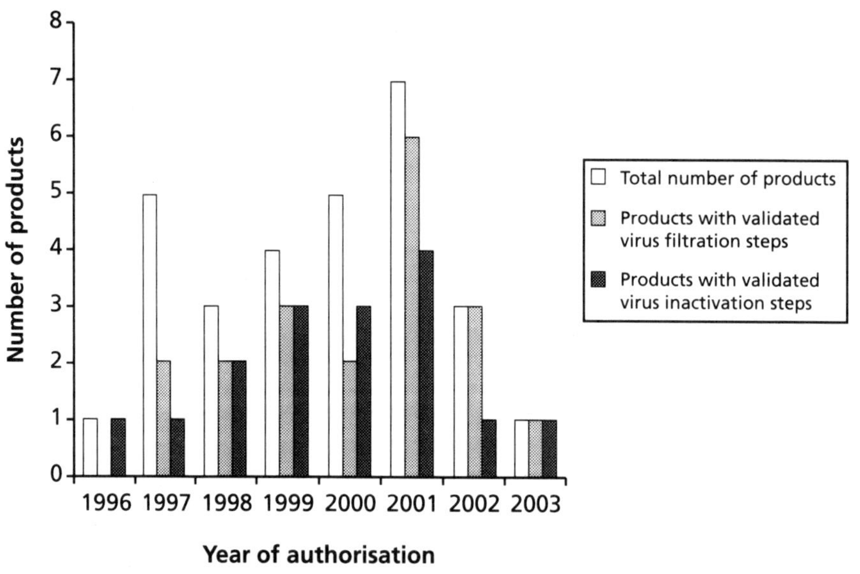

Fig. 1: Number of products with validated virus filtration or inactivation steps in the manufacturing process versus the year of authorisation.

A further analysis of the virus validation data showed that between two to six chromatographic resins are included in each virus validation study (most commonly two to four chromatographic steps are validated). Between three and six model viruses were used in the virus validation studies (in most cases four viruses were included). No trend was found when the number of resins or the number of model viruses was plotted against the year of authorisation.

A list of the most commonly used model viruses used in the virus validation studies was prepared (Table 2). A murine retrovirus was used in the virus validation studies of all 29 products.

The small non-enveloped viruses Reo3 and MVM were more commonly included in the virus validation studies of the products authorised after 2000.

Discussion

Despite the fact that the virus validation guidance is well known (having been in operation since 1997), virus safety issues arise frequently during assessment. On the positive side, major concerns have been identified during the assessment procedure for very few products only, i.e. insufficient virus safety assurance for the final products was shown. In all cases, these were products submitted for licensing before or soon after the Notes for Guidance on virus safety came into operation. Additional measures, such as additional inactivation or removal steps, more extensive virus safety testing or virus inactivation of the cell culture components had to be put in place before these products could be authorised.

Issues that are frequently raised for clarification include general concerns, concerns related to chromatography and concerns related to virus filtration. The general issues include: (i) reproducibility, reliability and specificity of the inactivation/removal; (ii) validity of the scaled-down validation procedure; (iii) clarification of process parameters; (iv) reduction capacity of the process for small

Table 2: Model viruses in virus validation studies.

Examples of model viruses given in the virus safety guidance documents[1]	**Model viruses actually used in the studies (frequency of use in brackets)**[2]
Large DNA viruses	
Herpesvirus	pseudorabies virus (PRV) (15/29); herpes simplex virus (HSV) (7/29)
Medium-to-large enveloped RNA viruses	
Parainfluenza or influenza virus; murine retrovirus; Sindbis virus	Murine retrovirus (29/29); parainfluenza virus type 3 (PI3) (5/29); bovine viral diarrhoea virus (BVDV) (4/29)
Small non-enveloped viruses	
SV40; poliovirus; animal parvovirus	Reovirus 3 (16/29); SV40 (13/29); polio virus (10/29); minute virus of mice (MVM) (11/29); bovine/porcine parvovirus (3/29)

1 CPMP/ICH/295/95 and CPMP/BWP/268/95

2 Only those model viruses used more than twice have been listed.

non-enveloped viruses, and (v) potential for virus contamination from animal- or human-derived culture media components. Questions raised on chromatography columns relate to (i) robustness/specificity; (ii) the effect of column re-use on virus partitioning, and (iii) column sanitisation and the mechanism of virus reduction (partitioning by the resin or inactivation by the elution buffer). Virus filtration often raises questions related to the robustness studies, virus aggregation (affecting the virus reduction) and integrity testing of the filters after use.

A survey was conducted to obtain a better insight on what has been carried out by the manufacturers of biotechnology products to ensure virus safety and to validate their manufacturing process. As the survey did not include a statistical analysis, the findings should be considered to be indicative only. The data set was analysed for: the inclusion of virus filtration steps and/or validated virus inactivation steps, the number of chromatographic steps validated and the number and type of model viruses used in the virus validation study. Where relevant, this information was correlated to the year of authorisation, in order to identify trends.

An initial finding was a shift in the manufacturing processes of the more recently authorised rDNA products and mAb to include virus filtration and specific virus inactivation steps. This reflects the increased importance that manufacturers are placing on the inclusion of more robust inactivation or removal steps in the manufacture of rDNA or mAbs. A second finding was that MVM is more frequently used in recent validation studies. This is probably a consequence of the more frequent use of virus filters, for which manufacturers tend to include small viruses (such as MVM). These small model viruses are not always included in the validation studies of the chromatographic columns. It probably also reflects the questions posed by regulators on the removal capacity of the purification process for small non-enveloped viruses and particular interest in MVM because there have been cases of adventitious contamination of fermentors with this virus. The number of model viruses used depends on the number of individual virus validation studies performed for a product; none of the single validation studies included more than four viruses.

Virus Safety Aspects for Plasma-Derived Medicinal Products

There are special aspects of plasma-derived medicinal products that require stringent attention to virus safety. In contrast to the well-defined and well-characterised cell lines used in biotechnology, the starting material is derived from a large number of blood/plasma donations. Viruses potentially present in donations can infect humans and in the past plasma-derived medicinal products have transmitted viruses, including HIV and hepatitis A, B and C viruses (HAV, HBV and HCV).

A variety of measures contribute to the virus safety of plasma-derived medicinal products: donor selection; screening of individual donations and plasma pools for virus markers; the inclusion of effective steps for the inactivation/removal of viruses in manufacturing processes; and Good Manufacturing Practice (GMP), including the segregation of material after virus inactivation/removal from untreated material to prevent cross-contamination. In addition, the presence of antibodies in immunoglobulin products contributes to safety against specific viruses such as HAV and B19 virus.

EU guidance on virus safety for plasma-derived medicinal products is provided in the guidelines on virus validation studies (CPMP/BWP/268/95) [1] and on plasma-derived medicinal products (CPMP/BWP/269/95) [4]. Specific advice on GMP for plasma-derived medicinal products is given in Annex 14 of the EU guide to GMP [7].

The current EU recommendations for viruses to be used in virus validation studies date from the mid 1990s. Enveloped viruses should include HIV-1, a model for HCV (e.g. BVDV, Sindbis or Yellow Fever virus) and an enveloped DNA virus (e.g. pseudorabies virus). Non-enveloped viruses should be selected to characterise the limits of the process. For coagulation factors, the viruses used should include HAV and a model for B19 virus (e.g. an animal parvovirus). In the case of immunoglobulins, viruses should be selected for which antibodies will not be present, as otherwise results will be confounded by the presence of antibodies.

The conference addressed a number of the current questions relating to virus validation studies including whether there is a practical model virus for HBV, the relative susceptibility of B19 and animal parvoviruses to inactivation/removal procedures, and the validation data needed in the case of newly identified/emerging viruses.

In 2000, the EMEA held a workshop on virus safety for plasma-derived medicinal products with a particular focus on non-enveloped viruses. A report and recommendations arising from the Workshop have been published on the EMEA website [8]. Current plasma-derived medicinal products have achieved excellent safety with respect to enveloped viruses including HIV, HBV and HCV. However, there are still limitations for non-enveloped viruses and improved inactivation/removal of non-enveloped viruses is an objective for plasma-derived medicinal products. This will improve the virus safety for known non-enveloped viruses and provide a first line of protection against other non-enveloped viruses (new, emerging, unidentified). Approaches for inactivation/removal of non-enveloped viruses were discussed in the session of the Forum on virus inactivation/removal technologies and in poster presentations. It is important that improvements in virus safety do not have adverse consequence on the efficacy and safety of products (e.g. protein modification and immunogenicity). Guidelines on the clinical investigation of the efficacy and safety of plasma-derived medicinal products are published on the EMEA website.

Recently CPMP, its Ad-Hoc Working Group on Blood Products (BPWG) and Biotechnology Working Party (BWP) have been considering how best to communicate information about the effectiveness of measures to achieve virus safety and any remaining potential risks. This has led to a change to the warning statements on transmissible agents in the product information for plasma-derived medicinal products. Instead of having a list of mandatory tests and a statement on effectiveness of manufacturing steps for virus inactivation/removal, the aim is to give clearer "bottom line" information about remaining risks. For example:

"The measures taken are considered effective for enveloped viruses such as HIV, HBV and HCV, and for the non-enveloped virus HAV. The measures taken may be of limited value against non-enveloped viruses such as parvovirus B19."

Following consultation with interested parties, final recommendations have been published on the EMEA website (CPMP/BPWG/BWP/561/03) [9].

In connection with this work on warning statements, the BWP is developing guidance on the risk assessment needed to support the warning statements and also for risk assessment in the case of emerging infections [10]. This follows the principle of weighing the potential virus input against the capacity of the manufacturing process to inactivate/remove viruses, in order to estimate the potential for virus contamination in the finished product.

West Nile virus (WNV) is a recently emerged infection in North America where it is now endemic. It is an enveloped virus belonging to the *Flaviviridae* family. In 2002 it became apparent that the virus could be transmitted by transfusion of blood components. However, no plasma-derived medicinal product has been implicated in WNV transmission. A CPMP Position Statement on WNV and plasma-derived medicinal products was published in 2003 [11]. On the basis of existing virus validation data, it was anticipated that inactivation/removal processes currently used in the manufacture of plasma-derived medicinal products would be effective for WNV. This anticipated effectiveness was demonstrated using WNV in several studies. The CPMP Position Statement concluded that the low levels and short period of viraemia and the effectiveness of inactivation/removal steps were reassuring. Further information on the incidence of infections, length and levels of viraemia has become available from the NAT testing of blood donations introduced in North America in 2003 and was presented during the Virus Safety Forum. The model virus approach to support the effectiveness of manufacturing steps for the inactivation/removal of WNV was also addressed.

This brief review illustrates how the virus safety of plasma-derived medicinal products is kept under review by CPMP and the BWP.

References

1 CPMP Note for Guidance on Virus validation studies: the design, contribution and interpretation of studies validating the inactivation and removal of viruses (CPMP/BWP/268/95), 14 February 1996.

2 ICH Q5A Note for guidance on quality of biotechnological products: viral safety evaluation of biotechnology products derived from cell lines of human or animal origin (CPMP/ICH/295/95), 4 March 1997.

3 Note for Guidance on the use of bovine serum in the manufacture of human biological medicinal products (CPMP/BWP/1793/02) of 18 June 2003.

4 CPMP Note for Guidance on plasma-derived medicinal products (CPMP/BWP/269/95 rev 3), 25 January 2001.

5 Note for guidance on minimising the risk of transmitting animal spongiform encephalopathy agents via human and veterinary product (EMEA/410/02 – rev. 2), October 2003. Official Journal of the European Union, C24, 28 January 2004 (2004/C24/03).

6 CPMP position paper on Creutzfeldt-Jakob Disease and Plasma-derived and Urine derived medicinal products (CPMP/BWP/2879/02), 20 February 2003.

7 Annex 14 to the EU Guide to Good Manufacturing Practice (ENTR/III/5717/99), 31 March 2000.

8 Report of EMEA workshop on viral safety of plasma-derived medicinal products with particular focus on non-enveloped viruses (CPMP/BWP/BPWG/4080/00) and Addendum: Conclusions and recommendations of the Biotechnology Working Party (BWP) and Ad-Hoc Working Group on Blood Products (BPWG) (CPMP/BWP/BPWG/93/01), 28 March 2001.

9 Note for Guidance on the Warning on transmissible agents in Summary of Product Characteristics (SPCs) and Package Leaflets for plasma-derived medicinal products (CPMP/BPWG/BWP/561/03), 22 October 2003.

10 Draft Note for Guidance on Assessing the risk for virus transmission – new Chapter 6 of the Note for Guidance on Plasma-derived medicinal products (CPMP/BWP/5180/03), 22 October 2003.

11 CPMP Position Statement on West Nile virus and plasma-derived medicinal products (CPMP/BWP/3752/03), 25 July 2003.

Dr. P. Celis, EMEA, 7 Westferry Circus, Canary Wharf, London E14 4 HB, UK.
E-mail: patrick.celis@emea.eu.int

Robertson J (ed): PDA/EMEA European Virus Safety Forum.
Dev Biol. Basel, Karger, 2004, vol 118, pp 11-15.

Viral Safety Evaluation of Plasma-Derived Therapeutic Products

M. Farshid

Center for Biologics Evaluation and Research, Food and Drug Administration*, Rockville, MD, USA

Abstract: The viral safety of plasma-derived products, in the U.S., is ensured by a complementary approach consisting of donor screening for high-risk activities or history, donor testing for relevant blood-borne viruses, and the inclusion of viral clearance steps in the manufacturing process. Thus, any manufacturing process for the production of plasma-derived products should be validated for its capacity to clear viruses. This validation is performed by using small-scale laboratory models that mimic the actual manufacturing process with regard to its critical operating parameters and performance characteristics. In addition, there are a number of validation components that are deemed essential for obtaining reliable viral validation data. These include: (i) validation of the small-scale model to establish its relevance to the actual process; (ii) choice of viruses in performing spiking experiments; (iii) reliability and validity of the assays used to determine viral titer in the test articles and (iv) calculation of total log reductions in the validated steps.

INTRODUCTION

Biopharmaceutical products that currently require viral validation studies can be loosely divided into three different categories: (i) products produced in cell culture, such as monoclonal antibodies and recombinant products. These products are obtained from well-defined and extensively tested starting materials, and altogether have an excellent viral safety record. (ii) Products derived from animals, such as anti-lymphocyte globulins produced in rabbits, antitoxins and antisera produced in horses and sheep, and various protein products produced in transgenic animals such as sheep and goats. Limited testing is done on the starting materials used in the production of these products, particularly products licensed several decades ago. (iii) Blood and blood products and other human-derived products. These are obtained from highly variable starting materials, which have in the past transmitted viruses. A number of complementary approaches are in place to reduce the potential

* The opinions and assertions expressed in this article are those of the author and are not to be construed as the policy of the U.S. Food and Drug Administration.

risk of viral contamination by these products and to increase their safety margins. This article focuses on human plasma-derived products. Products derived from animal plasma pose different risks from human plasma derivatives.

Addressing the risks that may be associated with plasma derivatives begins with the plasma from which the products are manufactured. These precautions include: (i) donor screening, which includes donor history assessment via questionnaires and interviews; (ii) maintaining donor deferral registries to prevent the use of units from unsuitable donors; (iii) donor testing, which includes serological testing for anti-HIV1/2, anti-HCV, and HbsAg, and nucleic acid testing (NAT) for HCV and HIV RNA; and (iv) the inclusion of effective viral clearance step(s) in the manufacturing process.

The aim of viral validation studies is to provide evidence that the production process will effectively inactivate and/or remove viruses, which could potentially be transmitted by the product, in the event that other strategies fail to detect such viruses. Viral validation studies also provide indirect evidence that the production process has some capacity to inactivate/remove potential viruses that are not included in the routine screening programme, and/or novel or yet undetermined viruses.

All processes which claim to clear virus should be fully validated by the manufacturer. Validation data should be specific to the product, the manufacturing process, and the virus. However, regardless of the setting in which the validation is performed, all viral validation studies share certain common features, which are briefly discussed below.

Validation of Scaled-down Laboratory Model

Obviously it is neither desirable nor technically feasible to do the validation studies in the actual manufacturing facility. Thus, a laboratory model, which is a scaled-down model of the manufacturing steps, is used for viral validation studies. The design of the laboratory model in any case should include all the critical parameters used at full scale and should adopt the production methods to the extent that is technically possible. This should include the control of parameters such as temperature, relative volume, linear flow rate, contact time, relative geometries, load and concentration. In addition, the actual performance of the scaled-down process should be identical to that of the actual manufacturing process, when measured by such parameters as relative capacity, yield, and purification. Therefore, the starting material for the process step (usually a process intermediate) should be of similar composition to that of the full-scale manufacturing process. This is often accomplished by sampling process material from a full-scale batch. Validation of the scaled-down model should involve multiple runs of the model, the results of which are compared to those from the full-scale process. Obviously, production methods are not always directly scaleable, and some methods may have to be modified. In such cases, these modifications should be justified and their impact should be addressed. In addition, validation studies using scaled-down systems should be designed to represent the worst-case conditions that could be encountered during full-scale production, i.e., those conditions under which the removal or inactivation of viruses would be expected to be least effective, but are still within the acceptable range for commercial manufacturing.

Spiking Experiments

Viral clearance validation studies usually take the form of "spiking" experiments, in which a large amount of a laboratory purified virus stock is added to the test article. Then the virus reduction factor for each step is determined. Appropriate controls should be included to ensure that the amounts of the virus do not change the performance of the scaled-down manufacturing step. Additional controls should also be included to confirm the viability of the indicator cells and the infectivity of the virus in the presence of manufacturing buffers. The viral reduction values obtained for each step are added to give a total log reduction for the entire manufacturing process. It should be noted that adding log reductions from different steps is acceptable only if the steps clear viruses by independent mechanisms. For example, viral reduction due to heat inactivation and that resulting from chemical treatment are additive; however, clearance values from two size-based filtration steps would not be considered additive. Each clearance step is usually validated as to its effectiveness and reliability independently of the others, although in principle, several steps could be validated in concert.

Assay Validation

For every step of manufacturing that is tested, there should be an efficient and reliable assay for the detection of each virus used. The titre of the viruses used in the validation studies are determined by using in vitro infectivity assays such as "plaque assays" or assays that measure "cytopathic effect," both of which are performed in cell culture. Biochemical assays such as antigen-based (immunoassays) or nucleic acid assays (e.g., PCR) may also be used, provided they are predictive of infectivity. Human pathogens such as hepatitis B virus (HBV) and hepatitis C virus (HCV) do not grow in cell culture, and in the past some studies have used animal models such as primates (HBV and HCV) or ducks (HBV) to validate the clearance of these viruses. However, such studies have become less frequent because of cost and availability, and because in most cases adequate alternative methods are available.

Choice of Viruses

Viruses that are used for validation studies fall into three categories: relevant viruses, specific model viruses, and non-specific model viruses. Relevant viruses are either the known pathogenic viruses of concern or of the same species or are viruses likely to contaminate the starting source materials or any other reagents used in the production process. When a relevant virus is not suitable for such studies (e.g., due to lack of an in vitro assay system), then specific model viruses are used as substitutes. An appropriate specific model virus is one that is closely related to the known or suspected virus (same genus or family) and has similar physical and chemical properties. Most validation studies are performed with specific model viruses sharing characteristics of the relevant human viruses. Relevant model viruses are used only for HIV and hepatitis A virus (HAV), for which appropriate in vitro systems are available. As mentioned earlier, the clearance of HBV and HCV has been validated in primate models, but such studies are not common today, and model viruses are used instead.

In addition to relevant and specific model viruses, additional viruses with differing structures and physicochemical behaviours should also be included in the studies, with the aim of obtaining results which can be used to demonstrate the capacity of the process to effect viral clearance in general. Here attention should be given to viruses that display a significant resistance to physical and/or chemical treatments. Table 1 shows model viruses that are currently used in viral validation studies of plasma-derived products.

Table 1: Model viruses for human blood-derived products.

Virus*	Model	Envelope/ Genome	Size (nm)	Resistance
HIV/HTLV	HIV-1	Yes/RNA	80-130	Low
HBV	DHBV	Yes/DNA	~ 40	Medium
HCV	BVDV	Yes/RNA	40-50	Medium
HAV	HAV	No/RNA	28-30	High
CMV	CMV/HSV/PRV	Yes/DNA	150-200	Low-Med
B19	PPV	No/DNA	18-26	Very high

* HIV= human immunodeficiency virus; HTLV= human T-cell leukaemia virus; HBV= hepatitis B virus; HCV = hepatitis C virus; HAV = hepatitis A virus; CMV = cytomegalovirus; B19= human parvovirus B19; DHBV= duck hepatitis B virus; HSV= herpes simplex virus; PRV = pseudorabies virus; PPV = porcine parvovirus.

Total Log Reduction Factor

In general, for a product to be safe, the process must remove or inactivate virus infectivity to a much greater extent than the level of virus infectivity that could potentially be present in the starting materials. For a process to have excess capacity with respect to virus reduction, it has been recommended that overall manufacturing processes should have at least two defined steps that are able to remove or inactivate viruses effectively and by different mechanisms. In addition, the inclusion of two complementary methods of virus inactivation and removal increases the spectrum of viruses that may be effectively cleared. At the risk of simplicity, for new products or products from new manufacturers, the rate and the extent of virus clearance should match those shown for products with excellent safety records.

Methods of Viral Clearance

Viral clearance is achieved by inactivation and/or removal steps in the manufacturing process. Common inactivation methodologies may be physical (e.g. heat treatment), chemical (e.g. solvent-detergent treatment) or physicochemical, which is a combination of the two (e.g. photochemical inactivation). The removal methodologies include precipitation (e.g. cold ethanol fractionation), chromatography (to effect separation by retaining either virus or active ingredient) and filtration (e.g.

nanofiltration, depth filtration). In all these cases, it is the responsibility of the manufacturer to validate the effectiveness of the method as well as to demonstrate that the validated operating parameters are met consistently during production.

Concluding Remarks

In evaluating viral validation results, the inherent limitations of viral validations studies should also be considered. These include the possibilities that: (i) the laboratory strains of viruses used in these studies may behave differently than native viruses; (ii) the source plasma or immune globulins may contain neutralizing antibodies, which could increase separation efficiency and inflate the viral removal value; (iii) the virus population may contain a fraction that is resistant to inactivation; (iv) the scaled-down process may be differ in performance from that of the full-scale process and give results that may differ to the actual situation; (v) total virus reduction may be overestimated because of repeated and similar process steps; and finally (vi) for certain validated steps, their ability to remove virus may vary after repeated use. Despite these limitations, the available data indicate that the technology of viral clearance is sufficiently developed to remove nearly all relevant enveloped viruses with a high degree of assurance. However, further development is needed to address the risk from small and highly resistant non-enveloped viruses such as human parvovirus B19, which could reach high titres in human plasma.

References

- Food and Drug Administration (FDA) Center For Biologics Evaluation and Research. Points to consider in the manufacture and testing of monoclonal antibody products for human use, 1997.
- International Conference for Harmonization (ICH). Text on validation of analytical procedures, vol. Q2A. ICH Technical Requirements for Registration of Pharmaceuticals for Human Use, Geneva Switzerland, 1994.
- International Conference for Harmonization (ICH). Validation of analytical procedures: methodology, vol. Q2B. ICH Technical Requirements for Registration of Pharmaceuticals for Human Use, Geneva Switzerland, 1996.
- International Conference for Harmonization (ICH). Viral safety evaluation of biotechnology products derived from cell lines of human or animal origin, vol. Q5A. ICH Technical Requirements for Registration of Pharmaceuticals for Human Use, Geneva Switzerland, 1998.
- Committee for Proprietary Medicinal Products (CPMP). The design, contribution and interpretation of studies validating the inactivation and removal of viruses. CPMP/BWP/268/95, 1996.

Dr. M. Farshid, CBER/FDA, Division of Hematology, HFM-345, Suite 200N, 1401 Rockville Pike, Rockville, MD 20852-1448, USA.
E-mail: farshidm@cber.fda.gov

Robertson J (ed): PDA/EMEA European Virus Safety Forum.
Dev Biol. Basel, Karger, 2004, vol 118, pp 17-29.

Current and Future Approaches to Ensure the Viral Safety of Biopharmaceuticals

K. Brorson[1], L. Norling[2], E. Hamilton[3], S. Lute[1], K. Lee[2], S. Curtis[2], Y. Xu[4]

1 Division of Monoclonal Antibodies, CDER/FDA, Bethesda, MD, USA
2 Department of Recovery Sciences, Genentech Inc., South San Francisco, CA, USA
3 Office of Vaccines Research and Review, CBER/FDA, Bethesda, MD, USA
4 World Wide Biopharmaceutical Development, GlaxoSmithKline, Inc., King of Prussia, PA, USA

The opinions expressed in this article are those of the authors and not necessarily those of the Food and Drug Administration or the U.S. Government.

Current Regulatory Approach to Viral Safety

Because of a continuing dialogue between industry and regulatory agencies around the world, robust measures to ensure the viral safety of biopharmaceuticals have evolved over the past two decades. A general approach specifically for biopharmaceuticals (e.g. recombinant proteins and monoclonal antibodies) has been accepted in several areas of the world and is described in regulatory documents ICH Q5a, Viral Safety Evaluation of Biotechnological Products Derived from Cell Lines of Human or Animal Origin, 1998 (http://www.ich.org/pdfICH/q5a.pdf), CBER's "Points to consider in the manufacture of Monoclonal Antibody products for human use", 1997 (http://www.fda.gov/cber/guidelines.htm), EMEA's "Note for guidance on virus validation studies: the design, contribution and interpretation of studies validating the inactivation and removal of viruses", 1996 (http://www.emea.eu.int/pdfs/human/bwp/026895en.pdf).

The viral safety approach for biopharmaceuticals consists of multiple levels of control at different stages of manufacturing [1-3]. It starts with rigorous screening of cell banks for bacteria, fungi, and mycoplasma as well as for viruses using several assays. The panel of virus assays that are used to screen the cell banks depends on the species of origin of the cell substrate (e.g. hamster versus mouse versus human). The goal is to manufacture biopharmaceuticals with cell lines demonstrated to be free of viruses. Practical limitations create exceptions to this goal, for example type C endogenous retrovirus particles are a known and often unavoidable by-product of mammalian cell culture. The second level of control is exercised at the cell culture harvest stage. Each harvest is screened for adventitious agents and type C particle levels are quantified in at least the first three validation cell culture batches.

As type C particles are a known, but quantifiable, upstream contaminant, the third level of control is a demonstration that the purification process can remove or inactivate them. This involves measuring the capacity of individual purification unit operations to clear model viruses, including murine retroviruses. These data are expressed as $\log_{10}$ reduction values (LRVs); the total clearance capacity of the process is the sum of the LRVs of each orthogonal unit operation in the process that is evaluated. To prevent the introduction of adventitious agents into facilities where biopharmaceuticals are made, these studies are performed in specialized laboratories outside the production area. The final levels of control are specific virus removal/inactivation steps added to the purification scheme, such as virus filtration, low pH treatment or solvent detergent inactivation.

Shared attributes of biopharmaceuticals

Common factors between the manufacturing of biopharmaceutical products and differences from other biologicals (e.g. plasma derived products, vaccines) have shaped a distinct overall strategy to ensure viral safety as well as particular tests. Except for biopharmaceuticals produced in bacterial cells, the manufacturing input material for recombinant biotechnology products are almost always cell culture harvests, often from rodent cell lines. These feedstocks are relatively uniform and controllable compared with those from other biologicals (vaccines, plasma products, etc.). In contrast, feedstocks for plasma-derived products are more heterogeneous and instances of contamination with pathogenic viruses have occurred in the past. While viral contamination of biotech feedstock materials is always possible, it is a rare event often resulting in dramatic cell-culture deviations such as loss of viability. The testing strategies for cell-bank screening are chosen based on the potential contaminants in the source material. Thus, while safety assurance for plasma-derived products uses screening for blood-borne human pathogens in each batch of plasma, the approach for recombinant products and monoclonal antibodies is simpler and more uniform. Cell banks are screened extensively for adventitious agents on a one-time basis, and harvests are screened with a smaller panel of assays. Endogenous type C retroviruses are an expected mammalian cell culture harvest contaminant and much is known about them scientifically. They are quantifiable using conventional (TEM) or new (Q-PCR) methods. Unlike the human pathogens potentially present in human plasma (e.g. HIV, HCV, etc.), it is unclear whether these particles are infectious in humans. In fact, Chinese hamster ovary cell (CHO) type C particles are regarded to be non-infectious even in cell lines [4]. Adventitious virus contaminations of commercial cell cultures have generally been rare and unique events. These events have been detected in the past by in vitro screens or impacts on cell culture, and have prompted extensive surveillance and decontamination activity.

Biopharmaceutical purification processes involve multiple steps, conferring another viral safety advantage to this product class. It has been evident for some time that many of these purification steps not only capture the product or remove extraneous proteins, but they also remove and/or inactivate viruses. Viruses, being complexes of proteins, nucleic acids and in some cases lipids, have specific surface charges, receptor affinities, chemical sensitivities and chromatographic behaviour just like smaller macromolecules or complexes. In addition, specific virus removal and/or inactivation steps can be added to biotech manufacturing schemes as long as they do not have an adverse impact on the product. The ability of bioprocess purification to reliably clear viruses has shaped the safety approach for clinical use and marketing. Evaluation of rodent retrovirus removal, using X-MuLV or a similar murine model retrovirus, is expected by regulatory authorities before clinical use of biopharmaceuticals derived from mammalian cell culture. To determine the safety factor of the product, the total clearance factor of the purification process is compared to the amount of type C endogenous retrovirus particles in a dose equivalent volume of harvest.

Evaluation or prediction of the process's removal capacity for adventitious virus is the more complex task of modelling the unknown. Because it is impossible to predict which virus will next contaminate a commercial bioreactor culture, these evaluation studies use a panel approach. Clearance of at least two to four additional viruses is evaluated before product licensing. The panels of viruses are chosen to maximize the diversity of virus structures and properties that are studied, e.g. RNA versus DNA virus, naked capsid versus enveloped virus, small versus large virus, physically resistant versus sensitive viruses.

While detailed guidance for planning and performing virus clearance studies is found in ICH Q5a; it is prudent to seek guidance from regulatory agencies to determine the scope and extent required for particular applications. For example, the panel of viruses that should be evaluated for clearance to support clinical trial use of a biopharmaceutical product may vary. Factors that have an impact on this issue include the nature of the product, its proposed indication, stage of trial (e.g. phase I versus II versus III), and the region of the world where the clinical trial will take place. It is particularly advisable to obtain regulatory guidance before use of biopharmaceuticals derived from human cell lines such as transfected PER.C6 cells or EBV transformed human cells.

Driven by the increasing size of the biopharmaceutical industry, advances in assay technology and scientific studies of biopharmaceutical manufacturing have accelerated in the past decade. A greater understanding of the properties of manufacturing unit operations and viruses that have an impact on viral flow, clearance and inactivation has resulted. This has created opportunities for modernization of viral safety practices based on scientifically founded principles. The remainder of this review will describe some of these opportunities.

Streamlined approaches

Certain unit operations have been repeatedly shown to remove or inactivate viruses in a robust manner [2]. Establishing the mechanism of clearance by these unit operations enhances the confidence that changing parameters unrelated to their mechanism of clearance will not have an impact on LRV. For example, low pH incubation inactivates murine retroviruses by driving pH-dependent chemical reactions such as irreversible conformational changes on viral surface proteins [5]. These chemical reactions are dependent on time, temperature and buffer pH, and unrelated factors such as concentrations of salts or non-viral proteins are predicted to have a less direct impact on the efficiency of this process. Thus, establishing a robust low pH unit operation should focus on time, temperature and buffer pH, rather than other variables.

By focusing on critical variables of unit operations, it is possible to adopt streamlined approaches to virus removal validation, particularly in the clinical phase of product development. Streamlined approaches include generic validation, bracketing and the combination of the two approaches. Generic validation was proposed in FDA's Points to Consider for Monoclonal Antibodies document (1997): "A Generic Clearance Study is one in which virus removal and inactivation is demonstrated for several steps in the purification process of a model antibody. These data may then be extrapolated to other antibodies following the same purification and virus removal/inactivation scheme as the model antibody." FDA has accepted generic data to support the IND use of monoclonal antibody products in cases where sufficient justification has been provided. FDA compares the unit operations of the model and follow-on products and an assessment is made of the degree of similarity between the critical operating parameters of the two processes. Generally, these critical parameters are the same or similar to those matched in scaled-down validation studies of large-scale operations, and should focus on those that are mechanistically most likely to have an impact on clearance.

FDA's Points to Consider Document for Monoclonal Antibodies also describes the streamlining approach of bracketing: "In some cases, sponsor may demonstrate virus removal/inactivation for a particular module at two different values of a given parameter (e.g. ionic strength, dwell time, temperature) and use any values of that parameter falling within this range." It is also feasible to validate viral clearance at "worst-case" set points and then justify operation below the "worst-case" set point. For example, validating clearance by a column packed at the minimum bed height likely to be used in operation would be "worst case" because it is known that chromatography performance and resolution improves with bed height. However, justification of this approach would require a detailed understanding of the mechanism of viral clearance by the unit operation.

Combining generic validation and bracketing can gain even more flexibility by defining operational ranges where viral clearance can be assured on a product-independent basis. Two examples of generic matrices have been published recently in the scientific literature; low pH murine retrovirus inactivation [5] and SV40 clearance by flow through mode anion exchange chromatography [6]. These publications are very recent and as yet it is unclear what their regulatory impact will be. It has yet to be established if they will be extensively cited in regulatory applications or if they will be accepted by regulatory agencies around the world. If widely adopted, bracketed generic validation will very probably afford considerable flexibility to IND sponsors. This flexibility will be particularly important for small firms or academic sponsors with limited resources which can be more productively applied to other aspects of product development. Bracketed generic validation will also guide manufacturers to choose operating set points which are within the ranges for robust viral clearance.

Q-PCR viral assays

After the invention of fluorogenic 5'nuclease PCR, new highly sensitive virus detection and quantification methods have been introduced into virus clearance studies [7-13]. These assays allow high throughput and rapid quantification of characterized viruses (e.g., where at least part of the genomic sequence has been resolved). These assays detect viral components (mainly nucleic acids, but some measure enzymes such as reverse transcriptase). Thus, they lend themselves to the evaluation of unit operations that separate virus from the product intermediate, but not necessarily to unit operations that inactivate virus. Virus removal unit operations include chromatography (e.g. anion exchange, protein A columns) and viral retentive filtration. Because they have large ranges (up to 6 $\log_{10}$) and great sensitivity (3 $\log_{10}$ more than infectivity assays), Q-PCR assays can accurately quantify the extent of viral reduction of unit operations that have high clearance capacities. Beyond validation studies Q-PCR assays can quantify endogenous retrovirus titres in cell culture harvests if the retroviruses released by the cell substrate have been appropriately characterized.

Two modes of Q-PCR virus assays have been described: genome specific [8-14] and reverse transcriptase (RT) specific [7,15,16] assays. To develop a genome-specific assay, primers are chosen from unique sequences in the viral genomes. In the first step of the assay, nucleic acids are extracted from viral particles in the test article. Assays for RNA viruses incorporate a cDNA synthesis step before PCR. Amplification & quantification of the viral DNA or cDNA is performed using fluorogenic 5'nuclease PCR.

Q-PERT (polymerase enhanced reverse transcriptase) assays are highly sensitive RT assays. They were originally developed for screening vaccine substrates and serum samples for retroviruses, but have been adapted at CBER for quantifying rodent retroviruses in models of bioprocessing [7]. In the first step of Q-PERT assays, RT in the test article synthesizes cDNA from a phage RNA template. This step is followed by an RNAse incubation to degrade the remaining phage- RNA. Finally, the phage-cDNA is amplified and quantified using Q-PCR technology; the number of cDNA copies of the phage correlates with the original RT activity in the test article.

The high throughput and rapid nature of Q-PCR assays represents an advance over traditional assays, lending themselves to several applications. Most obviously, they can replace the slower conventional assays used in validation studies; potentially lowering the overall cost in resources and time of these studies. Increased precision of Q-PCR assays may also lead to more accurate measurements of a unit operation's clearance capacity. Additionally, Q-PCR assays allow systematic scientific evaluations of unit operation parameters to determine which ones have an impact on viral safety [5,6,17,18]. Before the development of Q-PCR assays, these studies were very difficult or impossible due to high costs in resources and time. In contrast, Q-PCR assays are rapid and have a high throughput, allowing a large number of samples to be analysed in a timely manner. A recent study examining endogenous retrovirus expression in cell culture illustrates this point; type C particle loads were quantified using Q-PCR assays in >200 cell culture samples [18] This scope of work would have been prohibitive using conventional assays (e.g. TEM). As more studies are completed, a thorough understanding of the factors that affect endogenous retrovirus expression in cell culture and viral clearance during purification will be achieved. Manufacturers will be able to optimize unit operations for virus control *before* the final process is set and validated.

Q-PCR assays are new compared to the extensive experience with infectivity assays and TEM. Thus, questions have arisen about the regulatory acceptance of Q-PCR assays in validation studies. Specific guidance about applicability of Q-PCR assays is found in FDA's PTC for mAb 1997 (pg. 21): "... Assays which do not rely on infectivity, such as PCR-based assays, may be substituted *(for infectivity in clearance studies*), provided they have been validated for sensitivity and specificity, and that their results are correlated with those of infectivity assays." Thus, adequate validation is the key, at least for U.S. regulatory acceptance, and is most likely to be initially evaluated on an application-specific basis. Guidance on assay validation is found in ICH Q2A and B [19,20]. These documents list parameters that should be evaluated during the validation of quantitative assays: specificity, linearity, range, accuracy, precision, detection limit (LOD), quantification limit (LOQ), robustness, and system suitability. It is possible to evaluate each of these aspects of a Q-PCR assay, and an example has been published [15]. Beyond those listed in ICH Q2A and B, an additional aspect is described in the Points to Consider Document for Monoclonal Antibodies: "correlation" to conventional assays. "Correlation" is not defined precisely in the document. However, it can be presumed that it consists of accuracy and aspects of cross-validation to conventional assays. The interpretation of "correlation" should consider the fact that Q-PCR assays measure the total quantity of a viral component while infectivity assays measure infectious particles. Generally, only one in 100 to 1,000,000 particles in cell culture sample is infectious, depending on its nature and preparation method [15,21]. Thus, it is reasonable to presume a Q-PCR assay and an infectivity assay will measure a very different titre in a sample. While TEM quantifies total particle counts, they are less sensitive and precise than Q-PCR. Thus, correlation studies of Q-PCR and conventional methods should be interpreted carefully.

Because of the possibility of false signals, similar caution should be applied towards interpretation of clearance values measured by Q-PCR. False signals in Q-PERT assays can be caused by DNA contamination in the phage RNA template [15,22] or cellular polymerases in the test article [16,23-25]. However, it should be noted that most RT activity in murine hybridoma cell culture samples is associated with type C particles (Fig. 1a). Thus, while Q-PERT may overestimate

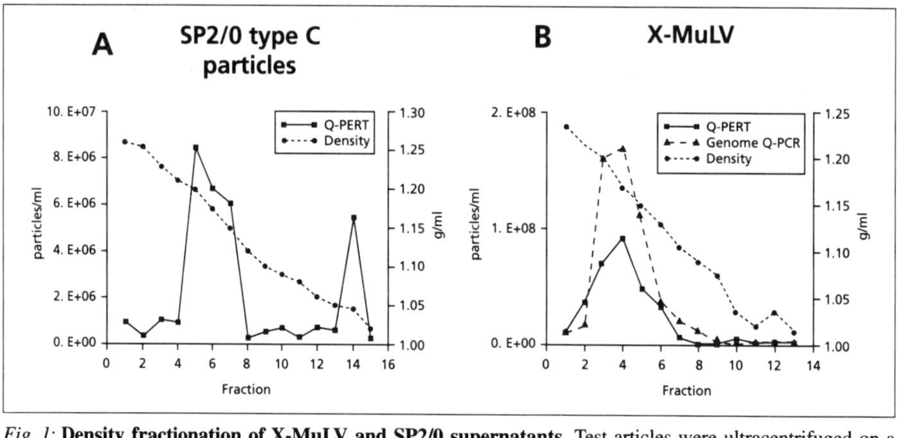

Fig. 1: **Density fractionation of X-MuLV and SP2/0 supernatants.** Test articles were ultracentrifuged on a 0-68% sucrose continuous gradient. RT levels and viral genome copy numbers in ≥13 fractions were measured by Q-PCR assays while densities were determined gravometrically. (a) RT activity and genomic RNA in X-MuLV preparations co-fractionate in a high-density fraction (1.17-1.20 g/ml). (b) A low-density peak (1.04 g/ml) contains ~17% of the total RT activity in SP2/0 supernatants, while the high-density peak corresponding to retroviral density (1.15-1.20 g/ml) contains 64% of the total RT activity.

type C particle quantities in murine hybridoma harvests, the type C particle component of the total signal detected by this assay is greater than the cellular polymerase component.

False signals in genomic assays can be amplified from nucleic acid fragments [12]. For example, even extensively purified virus can contain genomic and viral nucleic acid fragments. These fragments can co-amplify with nucleic acids contained within intact particles, causing an overestimation of the viral count. False-positive signals can also be amplified from viral nucleic acid fragments in animal-derived raw materials, potentially causing unnecessary batch rejections. Nucleic acid fragments are also problematic for filtration validation studies because they are smaller than viruses and can pass through viral-retentive filters. A signal from nucleic acid fragments in a virus-free filtrate will lead to an underestimated LRV for the filter. Conversely, nucleic acid fragments in a virus preparation used in spiking studies may lead to an overestimation of the input titre in the validation study and an overestimated LRV. A precaution that may minimize false signals is the incorporation of a nuclease step early in the Q-PCR assay (e.g. before capsid dissolution). Banding on density gradients can also be used to clean up virus preparations. Alternatively, measuring the nucleic acid content of density fractions from the viral preparation ultracentrifuged over sucrose density gradients can assess contamination with naked nucleic acid. In Figure 1b, this technique demonstrates the purity of an X-MuLV sample; almost all RT activity and genomic RNA resides in the density fraction consistent with the density of intact retroviral particles (e.g. 1.18-1.20 g/ml). A demonstration that validation studies are measuring clearance of intact virus, not free nucleic acids, is key for regulatory acceptance of Q-PCR methods.

Specificity considerations are also critical for the implementation of Q-PCR. Genome Q-PCR assay, should be considered to be specific for characterized viruses known to have the assay primer and probe sequences. Thus, genome Q-PCR assays can be considered for use in spike and removal validation studies, as homogeneous stocks of characterized virus are used. Q-PCR assays can also be considered for the enumeration of cell culture harvest type C particle loads as long as the primer and probe sequences are complementary to a majority of the retrovirus population produced by the substrate cell line [8]. This requires careful evaluation of cell culture [4,26-28], for example by sequencing a representative sample of particles made by the cell substrate. Such an evaluation done on the cell substrate used for transfection (e.g. the parental CHO or myeloma cell line), might be sufficient to support subsequent transfectants derived from them. Q-PERT assays may also prove suitable for enumerating type C particle loads given that they detect active RT, a component of all retroviruses [7,15]. To implement these assays, an assessment of the RT activity (in nano units) per virus particle should be made. Genome-specific Q-PCR is probably not suitable for other applications, such as the validation of virus inactivation unit operations (Low pH incubation, heat treatment, or solvent/detergent incubation). Given that these steps inactivate viruses without physically destroying them, Q-PCR assays specific for viral components may give false-positive signals even after complete viral inactivation. Newer Q-PCR- based assays such as infectivity PCR (iPCR) have been proposed as being sensitive and precise alternatives to traditional infectivity assays [29]. At this time, they are not in wide use. The Q-PCR format is also unlikely to prove suitable for screening for unknown viruses; no primer and probe set can be made sufficiently universal to detect all viruses, known and unknown.

Virus removal after extensive column resin re-use

A potential safety concern in biotechnology purification schemes that employ the re-use of column media, often for large numbers of chromatography runs, is a loss of the virus removal capacity of the chromatographic purification operation over time. Some controversy has arisen concerning how to demonstrate that products made with end-of-lifetime resins are safe. Because of this concern, an expectation has existed that end-of-lifetime column resins should be included in virus removal validation studies. This unwritten requirement has proved controversial because of the expense of these studies, the lack of written guidance, and the lack of a precedence of a clear safety problem posed by re-used column resins. Further, data from multiple applications have shown that LRV is stable as long as chromatography performance is maintained.

A recently proposed alternative approach is to define chromatography performance attributes that best predict virus clearance during the column lifetime and then monitor them in lieu of re-validation [17]. A recent set of small-scale experiments addressed this approach; protein A columns were cycled 150 to 460 times using murine hybridoma harvest concentrates, standard low pH elution buffers and different cleaning solutions. Protein A chromatography was chosen for study because it is probably the most common unit operation in mAb manufacturing. Viral clearance by protein A was also considered to be scientifically interesting because viruses flow through the column during mAb loading. These studies identified Ab step yield and breakthrough as performance attributes that decay before retrovirus LRV during the lifetime of the protein A media. Based on these data, it was concluded that an acceptable approach to protein A media lifetime validation was to perform small-scale viral validation studies on naïve resin only and then monitor Ab step yield and breakthrough during commercial production [17]. As a follow-up, we are implementing the matrix approach to evaluate protein A lifetime by comparing multiple commercial feedstocks and buffers, protein A media with different back-bone and cross-linking chemistries and extreme cleaning conditions. The follow-up studies are not only evaluating step yield and virus removal, but additional performance attributes such as clearance of host cell protein, DNA and media impurities, packing quality, and ligand leakage. By monitoring and comparing these performance attributes with each other, we hope to determine which are most predictive for media lifetime, and whether this varies depending on specific aspects of the media. The performance attributes identified to decay before viral LRV may then be used as indicators for the resin's ability to clear viruses over multiple uses during commercial production so as to eliminate the need to perform small-scale viral clearance studies with maximally cycled resin before product licence approval.

Viral clearance by re-used anion exchange (AEX) media is also of interest as AEX is the second most common chromatography unit operation in mAb manufacturing. In addition, anion exchange unit operations clear viruses by a mechanism distinct from protein A media and bind viruses more avidly than basic products such as mAbs. Thus, the observation that viral clearance by protein A chromatography is remarkably stable should not be assumed to translate automatically to AEX chromatography. To investigate the stability of AEX chromatography, Q-sepharose media was cycled in bind and elute mode 277 times with 2M NaOH cleaning (Fig. 2). Capture of the model protein, BSA, diminished by about 22%, indicative of some loss of the overall binding capacity of the column. Further evidence of column fouling or other degradation was detected by HETP analysis and frontal analysis of BSA break-through curves [30,31]. The decreased HETP values and asymmetry of the acetone peak indicate increased zone spreading or uneven flow of the acetone pulse as it traverses the column. This can be caused by degradation such as channel formation, resin fouling or loss of matrix support integrity. Increased early BSA breakthrough is indicative of a loss of the dynamic binding capacity of the column for the model protein relative to the total binding capacity. Loss of dynamic binding capacity can be caused by increased resistance to mass transfer, perhaps because of ligand loss or resin fouling. LRV of X-MuLV was stable, remaining consistently between 2-3 $\log_{10}$. In a separate experiment, the re-used resin was run in multi-virus spiking experiments [11] using a mAb as a model protein. Because a different model protein (a mAb) was run through the column in a different mode (flow-through), clearance of three viruses, SV40, X-MuLV and MVM, was higher than in the earlier experiment [6]. However, clearance of the three viruses was largely comparable between naive Q-sepharose media and the media that was cycled 277 times. While the LRV measured for X-MuLV and MVM were somewhat higher using the naive resins, viral clearance differences of <1 $\log_{10}$ are not considered significant from a regulatory standpoint [2]. Thus, it appears that decreased dynamic binding capacity or increased zone spreading detected by HETP or frontal analysis is most predictive of AEX column performance, and that viral clearance is stable even after moderate column decay. We are pursuing follow-on studies with AEX columns run in flow-through mode.

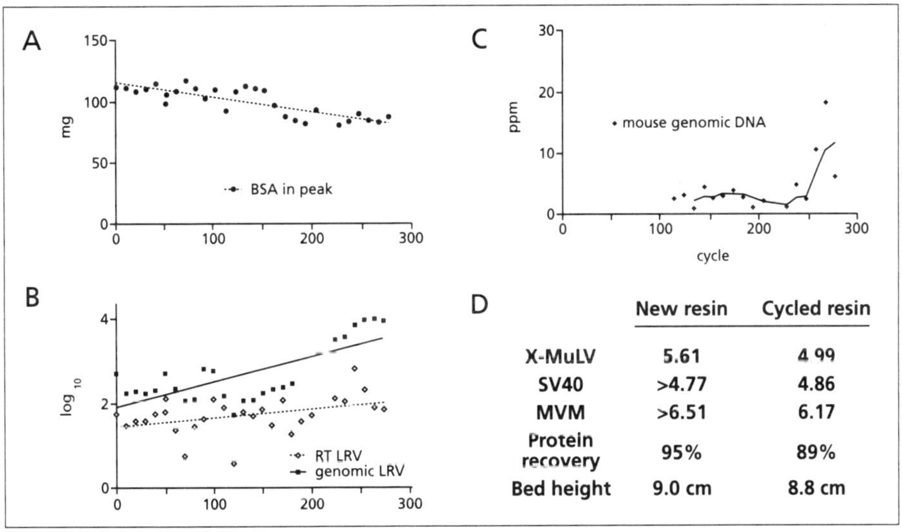

	New resin	Cycled resin
X-MuLV	5.61	4.99
SV40	>4.77	4.86
MVM	>6.51	6.17
Protein recovery	95%	89%
Bed height	9.0 cm	8.8 cm

Fig. 2: **Q sepharose Fast Flow (Amersham, Piscataway, NJ) column run through 277 bind & elute cycles with 2M NaOH cleaning.** The experiment was performed using a programmable AKTA explorer 100 system. The column diameter was 0.5 cm and the height was 11 cm. Each cycle consisted of the following sequential steps: a 7.5 column volume (CV) wash with cleaning buffer (2M NaOH); 10 CV wash with equilibration buffer (25 mM Tris pH 7.5, 10 mM NaCl); 5 CV BSA/DNA load in same buffer (10 mg/ml fraction V bovine serum albumin, 100 mg total; 500 ng/ml mouse genomic DNA was added to the BSA starting at cycle 103); 13.8 CV wash with equilibration buffer; 7.5 CV elution gradient to 250 mM NaCl; 2.5 CV gradient to 1M NaCl to strip column; 3 CV strip of

column in 1M NaCl; 5 CV wash in equilibration buffer. Every tenth cycle, starting at cycle one, the BSA/DNA load was spiked with 9 x 10^4 $TCID_{50}$/ml X-MuLV and the peak above 100 mAU during the elution step was collected. The flow rates varied by stage: 0.5 CV/min for cleaning, 1 CV/min for washing, 0.25 CV/min for loading, 1 CV/min for second wash, 0.65 CV/min for elution, stripping and washing. Every 10 cycles, buffer flow was paused and the column was stored for 60 minutes in 100 mM NaOH. (a) Total protein recovery in eluate peak measured by UV spectroscopy every ten cycles (diamonds). Dashed line is linear fit of protein recovery (b) X-MuLV clearance by column measured every ten cycles by RT specific Q-PCR assay (squares) and viral genome specific Q-PCR assay (diamonds). Solid and dashed lines are the respective linear fit of LRV. (c) Mouse genomic DNA present in eluate peak measured every ten cycles, starting at cycle 113, by a mouse B2 repetitive element Q-PCR assay (diamonds). Solid line is a three-point trend of parts DNA per million parts total protein in eluate peak. (d) The Q-sepharose resin was recovered after this experiment and reevaluated compared with naive resin (same lot) for clearance of MMV, X-MuLV and SV40 using a mAb model protein run in flow through mode (100 cm/hr flow rate, mAb loading at 70 mg/ml). The mAb flow-through and NaCl/Tris-based equilibration buffers were adjusted to 7.5 mS/cm, pH 8.0, as we have shown previously that under these conditions mAb flows through the column, while spike viruses bind to the QSFF resin [6]. (e) Measurement of zone spreading or height equivalent of a theoretical plate (HETP expressed as plates/column height in meters, N/m) at the start of the run using BSA as a model protein. HETP is generally considered to be a quantitative measure of column quality [30]. Each HETP measurement consisted of a 6.2 CV wash with dH_2O, 25 µl pulse of 2% acetone in dH_2O, followed by a 4.5 CV wash with dH_2O. The linear flow rate was 920 cm/hr. Asymmetry and HETP of the acetone A_{280} peak was calculated using the Unicorn software that runs the AKTA system. (f) Acetone peak after final cycle 277. (g) Frontal analysis of BSA breakthrough curve at cycle 62. Frontal analysis is a method for a quantitative assessment of the performance of chromatographic adsorption of proteins [30,31]. Each frontal analysis run consisted of a 7.5 CV wash with cleaning buffer at 0.5 CV/min; 5 CV wash with equilibration buffer at 1 CV/min; loading of 10 mg/ml BSA at 0.5 CV/min until the column was saturated (not to exceed 58 minutes). The A_{280} breakthrough curve generated upon BSA saturation was differentiated and measured for asymmetry using Unicorn software. (h) BSA breakthrough curve at cycle 237.

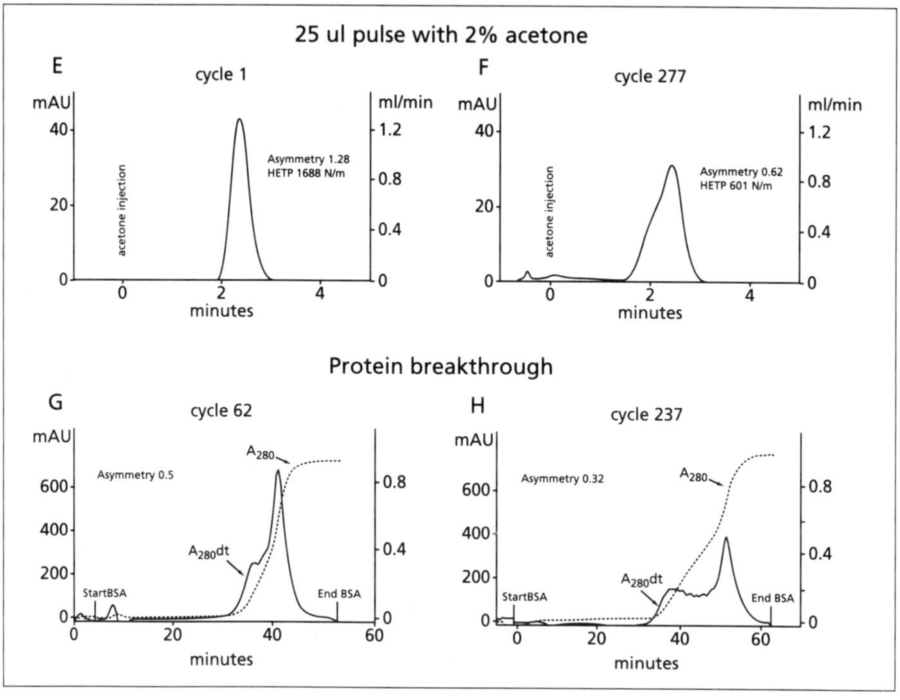

Viral retentive filter nomenclature

An exciting new development in the viral safety arena is PDA's effort to develop a common nomenclature for viral-retentive filters. Gail Sofer has described this effort in detail at this conference [32]. Briefly, the endeavour was prompted by inconsistency in the nomenclature systems for virus-removal filters. A PDA technical task force was convened to develop a common nomenclature and a standardized test method to be used by filter manufacturers for classifying and identifying viral-retentive filters. It was decided early in this process to focus on 'large virus' filters (e.g. Asahi Planova 35N, Pall DV50, Millipore Viresolve NFR) and that the best rating system will be based on retention of a bacteriophage, PR772. To gain consensus on this effort, it was stipulated that the purpose of the method was not to test filters at maximum or worst-case operating conditions or to compare filters from one filter manufacturer to the other. It was also agreed that the method was not intended to substitute for process validation to support regulatory compliance. Because virus filters from different manufacturers vary in construction and operation, the approach of the committee was to avoid promulgating a set uniform operating condition. Instead, the document specifies permissible ranges for relevant operating parameters based on industry practice, published literature, and filter manufacturer recommendations. The bracketed ranges are intended to allow testing of the filters at conditions recommended by filter manufacturers but at the same time are representative of actual commercial operations by end users. The document is almost complete and will be implemented on a prototype basis in 2004 at an FDA lab. The committee hopes that the 'large virus' filter method will be successful and will inspire a similar effort to classify 'small virus' filters.

Acknowledgements

We acknowledge Keith Webber (CDER/FDA), Mahmood Farshid (CBER/FDA), Qi Chen (Genentech), and Greg Blank (Genentech) for a critical review of our paper. We thank Billie Herzer and Mark Voisard of Amersham, Inc. for assistance with chromatography experiments.

References

1 Anonymous: Note for guidance on virus validation studies: the design, contribution and interpretation of studies validating the inactivation and removal of viruses, in C. f. P. M. Products (ed): European Agency for the Evaluation of Medicinal Products, Human Medicines Evaluation Unit, London, 1996.

2 Anonymous: Points to consider in the manufacture and testing of monoclonal antibody products for human use. Center For Biologics Evaluation and Research, Food and Drug Administration, U.S. Department of Health and Human Services, Rockville, MD, 1997.

3 Anonymous. Viral safety evaluation of biotechnology products derived from cell lines of human or animal origin, International Conference for Harmonization of Technical Requirements for Registration of Pharmaceuticals for Human Use, Geneva, Switzerland, 1998, vol Q5A.

4 Dinowitz M, Lie YS, Low MA, Lazar R, Fautz C, Potts B, Sernatinger J, Anderson K: Recent studies on retrovirus-like particles in Chinese hamster ovary cells, in Brown F, Esber EE, Williams M (eds): Continuous Cell Lines – An International Workshop on Current Issues. Dev Biol Stand. Basel, Karger, 1992, vol 76, pp 201-207.

5 Brorson K, Krejci S, Lee K, Hamilton E, Stein K, Xu Y: Bracketed generic inactivation of rodent retroviruses by low pH treatment for monoclonal antibodies and recombinant proteins. Biotechnol Bioeng 2003;82:321-329.

6 Curtis S, Lee K, Blank GS, Brorson K, Xu Y: Generic/matrix evaluation of SV40 clearance by anion exchange chromatography in flow-through mode. Biotechnol Bioeng 2003;84:179-186.

7 Brorson K, Swann PG, Lizzio E, Maudru T, Peden K, Stein KE: Use of a quantitative product-enhanced reverse transcriptase assay to monitor retrovirus levels in mAb cell-culture and downstream processing. Biotechnol Prog 2001;17:188-196.

8 de Wit C, Fautz C, Xu Y: Real-time quantitative PCR for retrovirus-like particle quantification in CHO cell culture. Biologicals 2000;28:137-148.

9 Shi L, Ho J, Norling LA, Roy M, Xu Y: A real time quantitative PCR-based method for the detection and quantification of simian virus 40. Biologicals 1999;27:241-252.

10 Shi L, Norling LA, Lau AS, Krejci S, Laney AJ, Xu Y: Real time quantitative PCR as a method to evaluate simian virus 40 removal during pharmaceutical protein purification. Biologicals 1999;27:253-262.

11 Valera CR, Chen JW, Xu Y: Application of multivirus spike approach for viral clearance evaluation. Biotechnol Bioeng 2003;84:714-722.

12 Xu Y, Brorson K: An Overview of Quantitative PCR Assays for Biologics Quality and Safety Evaluations, in Brown F, Lubiniecki AS (eds): Process Validation for Manufacturing of Biologics and Biotechnology Products. Dev Biol. Basel, Karger, 2003, vol 113, pp 89-98.

13 Zhan D, Roy M, Valera C, Cardenas J, Vennari J, Chen J, Liu S: Detection of minute virus of mice using real time quantitative PCR in assessment of virus clearance during the purification of Mammalian cell substrate derived biotherapeutics. Biologicals 2002;30:259-270.

14 Shi L, Chen Q, Norling LA, Lau AS, Krejci S, Xu Y: Real Time Quantitative PCR as a Method to Evaluate Xenotropic Murine Leukemia Virus Removal during Pharmaceutical Protein Purification. Biotechnol Bioeng 2004, in press.

15 Brorson K, Xu Y, Swann PG, Hamilton E, Mustafa M, de Wit C, Norling LA, Stein KE: Evaluation of a quantitative product-enhanced reverse transcriptase assay to monitor retrovirus in mAb cell-culture. Biologicals 2002;30:15-26.

16 Maudru T, Peden KW: Adaptation of the fluorogenic 5'-nuclease chemistry to a PCR-based reverse transcriptase assay. Biotechniques 1998;25:972-975.

17 Brorson K, Brown J, Hamilton E, Stein KE: Identification of protein A media performance attributes that can be monitored as surrogates for retrovirus clearance during extended re-use. J Chromatogr A 2003;989:155-163.

18 Brorson K, De Wit C, Hamilton E, Mustafa M, Swann PG, Kiss R, Taticek R, Polastri G, Stein KE, Xu Y: Impact of cell culture process changes on endogenous retrovirus expression. Biotechnol Bioeng 2002;80:257-267.

19 Anonymous: Text on validation of analytical procedures. International Conference for Harmonization of Technical Requirements for Registration of Pharmaceuticals for Human Use, Geneva, Switzerland, 1994, vol Q2A.

20 Anonymous: Validation of analytical procedures: methodology. International Conference for Harmonization of Technical Requirements for Registration of Pharmaceuticals for Human Use, Geneva, Switzerland, 1996, vol Q2B.

21 Losikoff AM, Poiley JA, Raineri R, Nelson RE, Hillesund T: Industrial experience with the detection of retroviruses, in Brown F, Esber EE, Williams M (eds): Continuous Cell Lines – An International Workshop on Current Issues. Dev Biol Stand. Basel, Karger, 1992, vol 76, pp 187-200.

22 Kothapalli R, Danyluck G, B.R., Loughran T: Problems associated with product enhancement reverse transcriptase assay using bacteriophage MS2 RNA as a template. J Virol Methods 2003;109:203-207.

23 Lovatt A, Black J, Galbraith D, Doherty I, Moran MW, Shepherd AJ, Griffen A, Bailey A, Wilson N, Smith KT: High throughput detection of retrovirus-associated reverse transcriptase using an improved fluorescent product enhanced reverse transcriptase assay and its comparison to conventional detection methods. J Virol Methods 1999;82:185-200.

24 Lugert R, Konig H, Kurth R, Tonjes RR: Specific suppression of false-positive signals in the product-enhanced reverse transcriptase assay. Biotechniques 1996;20:210-217.

25 Maudru T, Peden K: Elimination of background signals in a modified polymerase chain reaction-based reverse transcriptase assay. J Virol Methods 1997;66:247-261.

26 Anderson KP, Lie YS, Low MA, Williams SR, Wurm FM, Dinowitz M: Defective endogenous retrovirus-like sequences and particles of Chinese hamster ovary cells, in Horaud F, Brown F (eds): Virological Aspects of the Safety of Biological Products. Dev Biol Stand. Basel, Karger, 1991, vol 75, pp 123-132.

27 Anderson KP, Low MA, Lie YS, Keller GA, Dinowitz M: Endogenous origin of defective retroviruslike particles from a recombinant Chinese hamster ovary cell line. Virology 1991;181:305-311.

28 Lie YS, Penuel EM, Low MA, Nguyen TP, Mangahas JO, Anderson KP, Petropoulos CJ: Chinese hamster ovary cells contain transcriptionally active full-length type C proviruses. J Virol 1994;68:7840-7849.

29 Olszewski L: Infectivity PCR: an alternative for viral clearance and inactivation studies. PDA/FDA Viral clearance forum, Bethesda MD, 2001.

30 Giddings JC: Dynamics of Chromatography. Dekkar, New York, 1965.

31 Larson T, Davis J, Lam H, Cacia J: Use of process data to assess chromatographic performance in production scale protein purification columns. Biotechnol Prog 2003;19:485-492.

32 Sofer G: Nanofiltration- PDA virus filter task force status report. PDA/EMEA European Virus Safety Forum, Langen, Germany, 2003.

Dr. K. Brorson, Division of Monoclonal Antibodies, CDER/FDA, Building 29B, Rm 5E16, 29 Lincoln Dr., Bethesda, MD 20892, USA
E-mail: kurt.brorson@fda.hhs.gov

Robertson J (ed): PDA/EMEA European Virus Safety Forum.
Dev Biol. Basel, Karger, 2004, vol 118, pp 31-35.

An Industry Perspective on Plasma-Derived Products

I. von Hoegen

Plasma Protein Therapeutics Association, Brussels, Belgium

INTRODUCTION

It should be kept in mind that medicinal products prepared from human plasma are different from other medicinal products because they are manufactured from a natural source material. Human source plasma, as any natural source material used for the production of medicinal products, is different from standardized chemical source material from which classical pharmaceuticals are manufactured. Human source plasma has some intrinsic challenges in the form of blood-borne pathogens. Industry and regulators have learned lessons from the past and today plasma protein therapies have an excellent safety record and are among the most tightly regulated medicinal products, but the approaches vary all over the world, resulting in disharmonised regulatory requirements.

In addition to E.U. regulations, there are also national regulations in the current E.U. Member States that are not always identical and which must be respected. E.U. enlargement raises new challenges in achieving harmonisation as the various accession countries that joined the E.U. in May 2004 have their own regulations.

The issue of disharmonisation is not only restricted to the European Union; the United States of America, Japan and other parts of the world also have a role in alleviating unnecessary and expensive disharmonisation. Harmonisation is truly a global issue. However, it must be recognised that various efforts have been made to harmonise the different requirements, particularly within the E.U., but also between the E.U. and the U.S. The U.S. and the E.U. have recently agreed to exchange information on the quality, safety and efficacy of pharmaceutical products intended for human and animal use.

While steps are being taken to improve harmonisation further, industry, patients and regulators still face the impact of different requirements between countries. It is widely accepted that plasma products licensed in the major regions are of comparable quality and safety, but can often not be exchanged between countries due to incompatible regulations. One example – now harmonised – highlights the problems; from the mid 1990s, Belgium was not asking donors about residency in the U.K., which was required in Germany. As a consequence plasma from Belgium could no longer be imported into Germany. The lack of harmonization on the requirement for alanine aminotransferase (ALT) testing (a measure of abnormal liver activity and in the past seen as a surrogate marker of risk behaviour) within the E.U. still obstructs Marketing Authorizations in the European procedures of plasma derivatives produced from plasma not tested for ALT levels, although the CPMP, in its position statement issued in 1999 (Plasma-derived medicinal products: position paper on ALT testing (CPMP/BWP/386/99)), has concluded that there is no clear benefit from ALT testing for identified viruses as an indicator of liver cell damage.

The differences in requirements pose an enormous burden on the resources of companies as well as regulators to comply with national requirements. Furthermore, these additional burdens contribute significantly to increasing the costs of manufacturing plasma-derived medicinal products and the time needed for regulatory approval. At times this may result in shortages of some products in some countries which have actually occurred in the recent past.

Harmonisation Today

The significance of harmonisation cannot be stressed enough. It is obvious that harmonisation of regulatory requirements would necessitate a huge global effort but would be beneficial for all parties involved, particularly patients. This means greater harmonisation is a primary goal for the Plasma Protein Therapeutics Association (PPTA). The following areas would specifically benefit from harmonisation: development of ICH type guidelines on virus validation for plasma, pharmacopoeia requirements i.e. sensitivity and validation of NAT assays which cause a number of problems for manufacturers and regulators, serological testing and ALT testing, different donor selection criteria and requirements for storage of final products and lastly, freezing and storage conditions, which are different, for example, between the U.S. and Europe.

On the other hand positive examples should also be acknowledged. The Note for Guidance on plasma-derived medicinal products (CPMP/BWP/269/95 rev 3) has been internationally recognised. In 2003, PPTA submitted comments to a guideline on NAT testing to the Taiwanese Ministry of Health. As a result the Taiwanese authorities have implemented the proposed cut-off limit for Parvovirus B19 of 10^5IU/ml for the plasma pool. PPTA's own voluntary standard has set the same limit.

Currently, regulators and industry are provided with an opportunity for a harmonised approach with the "Note for Guidance on the Warning on Transmissible Agents in SPCs and Package Leaflets for Plasma Derived Medicinal Products" (CPMP/BPWG/BWP/561/03), which has been adopted by the CPMP Biotechnology Working Party in October 2003. The issue in question is the requirement for manufacturers to demonstrate that implemented measures are effective for the removal or inactivation of non-enveloped viruses.

Virus validation

A review of the current situation regarding regulatory guidance on effective measures for virus inactivation/removal identifies the CPMP "Note for guidance on virus validation studies: the design, contribution and interpretation of studies validating the inactivation and removal of viruses" (CPMP/BWP/268/95). This document is one of the examples where a guideline has been adopted as an ICH guideline for biotech products and becomes relevant beyond the European Union. In the guideline, three principal complimentary approaches for calculation of the safety factor for a product of a given dose are listed. First is the selection and testing of source material for the absence of detectable viruses. The PPTA voluntary standard for Parvovirus B19 falls under this heading. Second is the testing of the product at appropriate stages of production for freedom from detectable viruses. In the case of plasma products, the plasma pool represents the control point for the safety of the starting material complemented by testing the capacity of the production process to remove or inactivate viruses. This is done by virus validation studies, where wide ranges of viruses with known physico-chemical properties are evaluated. The overall virus reduction for the manufacturing process is quantitatively evaluated and indirect evidence that the process will inactivate and/or remove novel or emerging viruses is provided, based on the use of relevant model viruses with a wide range of physico-chemical properties. In the Note for Guidance, the effectiveness of virus validation studies is limited to virus reduction by manufacturing processes, as NAT testing is currently outside the scope of this guideline and does not address how to demonstrate the effectiveness of measures for non-enveloped viruses. The CPMP has acknowledged the need for more specific guidance concerning the definitions and criteria for effective measures for non-enveloped viruses and has published the draft "Note for Guidance on assessing the risk for virus transmission – new chapter 6 of the Note for guidance on plasma derived medicinal products" (CPMP/BWP/269/95 rev 3) for comments. PPTA members have years of experience in working with all the different measures. On numerous occasions they have shared their experience with regulators. No "magic bullet" exists against non-enveloped viruses, therefore continuous discussions between authorities and industry experts are vital for developing and updating specific guidance to reflect the current state-of-the-art.

PPTA and its members believe that regulatory guidance agreed between industry and authorities will ensure regulatory compliance and encourage manufacturers to implement additional measures. However, more importantly, it must be recognised that regulatory guidance on effective measures against non-enveloped viruses is not an issue restricted to the European Union, but is of global relevance and therefore input from non-E.U. regulatory authorities should be taken into account whenever possible.

Emerging pathogens and the model virus approach

Emerging pathogens represent another area of challenge, where harmonised approaches are needed. PPTA member companies have reacted to recent challenges and have demonstrated that virus inactivation/removal methods already implemented into their manufacturing processes are effective against a variety of viruses such as West Nile Virus (WNV), vaccinia virus, monkey pox virus or Severe Acute Respiratory Syndrome (SARS). These assessments are based on peer-reviewed literature and/or verified by the use of model viruses, and in many cases also by using the actual virus itself. On numerous occasions, PPTA members have presented their data. Experience has shown that the risk of transmission of emerging pathogens is reduced for plasma-derived therapies because of established and validated virus inactivation and removal methods during manufacture, which cover a wide range of enveloped and non-enveloped viruses. It is required that manufacturing processes for biological products must demonstrate that they provide effective inactivation and/or removal of enveloped viruses. WNV is an enveloped virus sharing the common feature of a membrane envelope with other enveloped viruses, in particular with HCV model viruses where PPTA members have an abundance of experience.

From PPTA members' experiences, the model virus approach has proved to be an effective tool to deal with emerging pathogens. The use of model viruses is an inherent characteristic of virus validation studies and, as stated in the note for guidance on virus validation studies, any virus used in a validation study is actually a model virus. Model viruses are effective surrogates for closely related viruses and it has been shown that different enveloped viruses are similarly susceptible to different inactivation methods. The safety record of PPTA members' plasma-derived therapies validates laboratory clearance studies using models of clinically relevant viruses. Model virus studies predict that WNV, vaccinia virus, SARS-associated virus and monkey pox virus will be inactivated by manufacturing processes for plasma-derived therapies. Looking at recent studies performed in the U.S., it can be expected that donation loads of WNV in plasma pools will be below the limit of detection for current test strategies (1.5 x 10^5 cps/ml max /donation).

Responding to the Challenge

It is difficult to determine what the next "virus du jour" will be and several factors will be responsible: environmental conditions promoting propagation and distribution (e.g. WNV), the development of a new homotropic virus from a harmless animal virus (e.g. SARS), or the emergence of an at-the-time previously unknown virus, such as HIV two decades ago. On top of all these factors, global travellers will contribute to the worldwide dispersion of viruses. As history shows, our seafaring ancestors wiped out entire populations through the transmission of pathogens against which the native population had no immunity. The good news is that today we have means to respond to these challenges, but we have to make sure that we are doing the right thing.

To ensure a successful response to these challenges, we need to learn from our past experiences, one aspect being to accept the model virus approach. A global collaboration must be developed to respond to challenges posed by emerging pathogens, for example in the form of workshops on emerging pathogens with participation from regulators, industry, scientists and patients. A globally accepted emerging pathogen risk assessment matrix to provide a common basis is urgently needed.

The implementation of disharmonised donor selection or exclusion criteria must be avoided. The performance of validation studies for each "virus du jour" must not be required as it may be time-consuming, costly and wasteful. The value of the additional information must be questioned and weighed against the potential bio-safety level of an emerging pathogen which not only implicates the safety of the laboratory personnel but also the world-wide laboratory capacity that is available to perform studies at high bio-safety levels.

In conclusion, while globalization imposes new challenges on plasma protein therapies, it also provides opportunities for greater harmonisation. The need for regulatory guidance that takes the global perspective and harmonization into account is imperative. This will ensure worldwide regulatory compliance, will reduce unnecessary burdens and costs and, most importantly, increase product availability to patients.

Dr. I. von Hoegen, Plasma Protein Therapeutics Association, Boulevard Brand Whitlock 114/5, B-1200 Brussels, Belgium.
E-mail: ivonhoegen@pptaglobal.be

Robertson J (ed): PDA/EMEA European Virus Safety Forum.
Dev Biol. Basel, Karger, 2004, vol 118, pp 37-44.

Industry Perspective for Biotech Products

A. Möritz

Novartis Pharma AG, Technical Research & Development Quality Assurance, Basel, Switzerland

Key words: Biologicals, regulatory, guideline, cell-derived rDNA product, requirement, development.

Abstract: This presentation, held at the Joint PDA/EMEA Virus Safety Forum, provided an industry perspective on virus safety for biotech products. After discussion of the various groups of biologicals for human use, the presentation focused on a specific group of biologicals, the cell-derived rDNA products. The current authority guidelines of this group of well-defined biotech products were discussed, and a representative case study addressing virus safety aspects was presented. Questions for discussion were raised which illustrated the need for discussion between regulators and industry concerning updated guidance for this group of biologicals.

Biotech Products for Human Use

Biotech Products for human use can be classified in various ways, and Table 1 provides a proposal for a historical classification. The products named under "classical products" have been known for quite a number of years, and there are many molecules of this class which have already received marketing authorisation, or are in the licensing process.

The "modern products" group has been known for some time now, but this group has received only a limited number of marketing authorisations so far, mainly due to side effects that were reported in clinical trials.

Members of the group of "dernier cri" products are currently in clinical development, without marketing authorisations granted so far.

The cell-derived rDNA products, and especially the subgroup of monoclonal antibodies, were chosen as the main topic for the presentation, as they represent a group of products for which final regulatory guidance on virus safety has been available for some years although, historically, they are the youngest of the group of classical products. They are also a group for which many marketing authorisations have been granted since the availability of virus safety guidance.

Table 1: Historical classification of Biotech products.

"Classical products"

- Animal Sera and Immunoglobulins
- Products extracted from human/animal tissues/fluids
- Plasma Derivatives (special group of above)
- Vaccines
- Cell-derived Products including rDNA Products
  - Bacterial Fermentation
  - Yeast Fermentation
  - Mammalian Cell Culture
    - **Monoclonal Antibodies**
    - Cytokines
    - Fusion Proteins

"Modern products"

- Human Tissue-Based Products
- Human Cell Therapy
- Xenogeneic Cell Therapy
- Gene Therapy

"Dernier cri"

- Products from Transgenic Plants
- Products from Transgenic Animals

Regulatory Virus Safety Guidance Documents for Cell-Derived rDNA Products

The virus safety guidance available for cell-derived rDNA products is summarised in Table 2. The guidelines were classified by region, starting with the ICH guidelines, published in 1997, whose applicability covers the regions of the European Union (E.U.), the U.S. and Japan.

The WHO Technical Report Series is the regulatory source for the many more members of the WHO worldwide. Annex 5 of this report series, however, mentions as a reference list additional WHO guidance that was published in previous years. Although this referencing is very useful historically, it is not clear which virus safety guidance would apply to a given product.

In the E.U., a guideline on virus safety was published in 1995. It covers a wider range of products than the cell-derived rDNA products which are the sole topic of the ICH documents. The only additional document on virus safety that was published recently in the E.U. is the guidance on the use of bovine serum. The E.U. TSE Guideline – currently under revision – is mentioned in the E.U. list as the major additional guidance on adventitious agent safety that has been published since the 1995 guideline. As with WHO, also for the E.U., additional guidance documents dating from before the E.U. and ICH guidelines are published on the E.U. Commission's homepage.

In the U.S., the only guidance provided in addition to the ICH documents is on monoclonal antibodies. The two guidelines do not deal with virus safety alone, but also mention aspects of the development of monoclonals, such as pre-clinical and clinical aspects. The original guideline, published in 1997, is also the only guidance that provides some information on virus safety considerations in development. This topic has not yet been covered by any ICH, WHO or E.U. guidance. Concerning TSE safety for Biotech products, the FDA covered the issue mainly with letters to manufacturers; the most recent reference is mentioned in Table 2. The FDA CBER guidance document page also mentions additional guidelines from the years before the ICH guidelines were published.

Table 2: Virus safety guidelines for cell-derived rDNA Products.

- ICH
  - Q5A: Viral Safety Evaluation of Biotechnology Products Derived from Cell Lines of Human or Animal Origin, 4th March 1997 (http://www.ich.org/pdfICH/q5a.pdf)
  - Q5D: Derivation and Characterisation of Cell Substrates Used for Production of Biotechnological/Biological Products, 16th July 1997 (http://www.ich.org/pdfICH/q5d.pdf)

- WHO
  - WHO Technical Report Series 878, 1998 (adopted 1996), Annex 1 Requirements for the use of animal cells as in vitro substrates for the production of biologicals (http://whqlibdoc.who.int/trs/WHO_TRS_878.pdf)
  - WHO Technical Report Series 878, Annex 5 (same document) mentions additional guidance

- E.U.
  - CPMP/BWP/268/95: Note for Guidance on Virus Validation Studies: The design, contribution and interpretation of studies validating the inactivation and removal of viruses (http://www.emea.eu.int/pdfs/human/bwp/026895en.pdf)
  - DRAFT (Rev. 2) EMEA/410/01: Note for Guidance on Minimising the Risk of transmitting Animal Spongiform Encephalopathies via human or veterinary medicinal products (http://pharmacos.eudra.org/F2/pharmacos/docs/guidRtTSE%20Note%20for%20Guidance%20revision%20Dec.pdf)
  - CPMP/BWP/1793/02: Note for Guidance on the use of Bovine Serum in the Manufacture of Human Biological Medicinal Products (http://www.emea.eu.int/pdfs/human/bwp/179302en.pdf)
  - Additional guidelines are listed under: http://pharmacos.eudra.org/F2/eudralex/vol-3/home.htm
- U.S. – Guidance which mentions virus safety amongst other points
  - CBER Points to Consider in the Manufacture and Testing of Monoclonal Antibody Products for Human Use, 28th February 1997 (http://www.fda.gov/cber/gdlns/ptc_mab.pdf)
  - CBER/CDER Guidance for Industry: Monoclonal Antibodies Used as Reagents in Drug Manufacturing, March 2001 (http://www.fda.gov/cber/gdlns/mab032901.pdf)
  - BSE (latest publication): Letter to Manufacturers 19th - April 2000 (http://www.fda.gov/cber/ltr/bse041900.pdf)
  - Additional guidelines are listed under: (http://www.fda.gov/cber/guidelines.htm)

Abbreviations			
CBER	Center for Biologics Evaluation and Research	MAP	Murine Antibody Production
CDER	Center for Drug Evaluation and Research	MCB	Master Cell Bank
EMEA	European Medicines Evaluation Agency	PDA	Parenteral Drug Association
E.U.	European Union	Q-PCR	Quantitative Polymerase Chain Reaction
FCS	Foetal Calf Serum	rDNA	Recombinant Desoxyribonucleic Acid
FDA	Food and Drug Administration	TSE	Transmissible Spongiform Encephalopathy
HAP	Hamster Antibody Production	U.S.	United States
ICH	International Conference on Harmonisation	WCB	Working Cell Bank

In summary, most guidelines on virus safety for cell-derived rDNA products were published during the years 1993-97, with 1997/8 being the years of the ICH and WHO guidance publications. The E.U. and U.S. have published additional guidelines, which deal with cell-derived rDNA products, with a more general scope in the E.U. and a more focused scope in the U.S. After 1998, hardly any further guidance was set up by the regulatory authorities for this group of biologicals. So it appears that the ICH process brought about harmonisation to a certain extent, but not full harmonization, as additional guidance was deemed necessary by both authorities. All authorities still have additional guidance documents on their Internet pages. This is quite confusing, as these guidelines in some cases mention requirements additional to the documents which are the more recent, and therefore supposedly, the applicable ones.

The following questions therefore arise:

- Which guidelines would apply for a company wishing to develop a product for global submissions?
- Which guidelines could be regarded as outdated?

The following proposal was provided:

The ICH guidelines are the first choice for consultation regarding virus safety requirements. The E.U. and U.S. guidelines should be consulted for additional points that may not be covered, or are not sufficiently clear in the ICH documents. The WHO guidelines may prove to be a very useful source to provide an overview on the topic.

However, the following questions remain for further discussion:

- Is further harmonisation (or "face-lifting") of the available guidance foreseen?
- Is a guidance document on development in virus safety needed?

The Case Study – An Example from Research to Market

A monoclonal antibody project was chosen as an example for discussing current virus safety testing strategies. Table 3 presents the general features of the monoclonal discussed.

Table 3: Features of the monoclonal antibody discussed in the case study.

- The molecule's origin
  - Rodent cell line (with endogenous retrovirus) transfected with an rDNA construct (e.g., Sp2/0, NSO, CHO)
  - Fully human antibody
- Indication
  Chronic disease, with new mode of action compared to established treatment

The choice of the case study was based on final virus safety guidance available for this group of biologicals, the transfer of this group within the FDA from CBER to CDER, and the availability of a draft monograph of the European Pharmacopoeia for monoclonal antibodies.

All the above points show that at least the E.U. and U.S. regulators appear to be starting to see monoclonal antibodies as a kind of "commodity" – the reason also being that so many biologicals with monoclonals as the active pharmaceutical ingredient have already been licensed.

Furthermore, monoclonal antibodies as mammalian cell-culture derived rDNA products were among the first of their group that reached the market, with now well-defined cell lines that were used for transfection, such as CHO, NSO or Sp2/0. So far, for monoclonals, there was no drug safety incident reported related to virus contamination.

In summary, a review of virus testing strategies appears feasible for monoclonal antibodies manufactured from well-defined cell lines. There also exists an attractive previous example of successfully revised regulatory requirements: At the time, residual host cell DNA was seen as a major risk connected with cell-derived rDNA biologicals. This requirement was revised when sufficient experience was available that there is actually little risk. A review of virus testing strategies may yield the same result.

Transfer from Research to Development

The transfer of a project from research to development usually includes a handover procedure, which includes a compilation of the cell line history for any cell-culture derived molecule. This history should include information about the TSE safety of the cell line, with justification of the previous use of materials of bovine origin. As the cell lines discussed here are established lines which were originally transferred from university laboratories via cell culture collection institutions to a GMP environment, such safety considerations should be confined to the stages on which a company could control the culturing of the cells. Previous stages can only be covered by descriptive statements available from the scientific literature.

Concerning FCS that may be used for transfection, the new E.U. guideline on bovine serum needs to be taken into account, which also provides advice on when it is to be applied. In this respect, it is a point for discussion why serum that is used for the establishment of cell banks could not be excluded from the scope of the guidance, as thorough virus safety testing is to be performed on those banks.

Virus Safety Testing at the Start of Development

Cell line testing

Preparation of material for first-in-man studies is usually based on an MCB only, the WCB being prepared at a later stage of development. Virus testing is therefore done on the MCB, the unprocessed bulk and the "cells at the limit of in vitro cell age or beyond" (sometimes also called "end-of-production cells" or "extended cell bank", depending on which stage is chosen to perform the testing; the term provided here is the one used in the pertinent ICH guideline, which covers all options).

Here, a point for discussion is whether the testing of adventitious agents of the MCB for the case study as described could be suspended, if:

- the parental cell line was extensively characterised with regard to adventitious agents and
- only thoroughly tested – if any – animal or human origin material is used for establishment of the MCB. An example of such a thoroughly tested material would for instance be licensed Human Serum Albumin, which itself is a medicinal product undergoing numerous virus safety tests to be cleared by regulatory authorities.

In this scenario, virus safety testing of the MCB for first-in-man studies could be confined to the endogenous retrovirus, the most relevant virus for the well-defined cell lines discussed here.

Concerning the testing of an MCB for adventitious agents, there is also another aspect to be taken into account: is it still ethical to perform MAP/HAP tests on every MCB? Authorities (and to some extent, also contract laboratories) have access to a broad database of licensed monoclonal antibody products for which these tests were performed – were any of these tests ever positive? If this was not the case, it needs to be discussed whether the tests could be omitted when established, well-characterised parental cell lines as mentioned are used for transfection. And, if these tests were not to be omitted, are there suitable surrogate tests available to avoid so many animal tests as currently needs to be performed?

Downstream processing

Purification of a monoclonal antibody from unprocessed bulk comprises several standard elements: Protein A Affinity Chromatography, a Low pH step, a Nanofiltration step, and a Sizing and/or Anion Exchange Chromatography step.

Concerning virus validation studies of such a downstream process for a first-into-man study, it is currently unclear in general to what extent a virus validation using model viruses would be necessary, as only very limited regulatory guidance exists. From experience, it appears acceptable for regulatory authorities in the E.U. and the U.S. to only perform validation of the endogenous retrovirus inactivation and removal in singlicates for Phase I and II studies, provided well-defined cell lines and molecules as presented by the case study are used. The U.S. guideline on monoclonal antibodies even allows for exceptions from this for serious or immediately life-threatening conditions. There appears, however, to be a need for more guidance at least in the E.U., as harmonised requirements for the manufacture of Investigational Medicinal Products will enter into force in 2004. One of these requirements (Annex 13 No. 17, EU Guide to GMP) reads as follows:

"Likewise, when required, virus inactivation/removal and that of other impurities of biological origin should be demonstrated, to assure the safety of biotechnologically derived products, by following the scientific principles and techniques defined in the available guidance in this area."

As there is currently no guidance available on the E.U. level, a non-harmonised approach could lead to different regulatory authority opinions and practice within the E.U., thus obstructing development of innovative biological products.

Another step of downstream processing, nanofiltration, has been shown to be very effective in the removal of most viruses. Therefore, discussion is needed as to whether validation of this step for endogenous retrovirus removal is still necessary for every monoclonal antibody during early development.

Virus Safety Testing During Further Development

Aspects concerning cell-line testing

At later stages in the development of a monoclonal antibody, for example, before Phase III, the "final" MCB as well as a WCB are established. Basically, the same discussion concerning MCB testing as for early development described above could be possible.

Another interesting aspect concerning MCB testing would be whether quantification of the endogenous retrovirus could also be performed by Q-PCR and what the requirements for cross-validation with negative stain electron microscopy would be. Currently, Q-PCR is only used for unprocessed bulk testing and downstream process virus validation testing.

As mentioned by the pertinent ICH guideline, WCB testing is suspended in favour of testing of the "Cells at the limit of in vitro cell age or beyond". Concerning testing of the unprocessed bulk, clarification is needed whether in vivo testing of one batch is still deemed necessary by the FDA, as mentioned in their guidance on monoclonal antibodies.

Filing, Approval, Post-approval Development Stages

Aspects concerning downstream processing

At the filing/approval stage, the requirements for demonstrating virus safety of a downstream process are mostly defined for any biological, as a full virus validation study using three to four model viruses is to be performed on the final purification process. However, concerning monoclonal antibodies, where several, well-defined process steps are used, the concept of generic or modular clearance studies as outlined in the U.S. Guideline on Monoclonals needs more consideration. Could this concept be acceptable at least for some aspects of the final purification process, for example, column cleaning and regeneration procedures, which are very similar, if not identical, for monoclonal antibody processes? There is no doubt that the concept itself could also prove useful if applied during development (cf. the respective reports presented in this volume).

One post-approval aspect which is usually to be addressed is the virus safety testing of chromatographic resins at or past their defined lifetime, as well as the use of new resins. Discussion is needed whether virus validation studies are still needed for these cases if the performance of the column is well documented.

Conclusions

The virus safety guidelines applicable to cell-derived rDNA products were discussed, with the conclusion that a review and, if possible, further harmonisation of the existing guidance documents available should be taken into consideration by the regulatory authorities. It should also be discussed whether a guidance on virus safety for cell-derived rDNA products in development might add a useful element of guidance to industry.

The case study presented demonstrated that there are strong arguments in favour of a re-thinking of the current virus testing strategies for well-defined cell-derived rDNA products such as monoclonal antibodies.

Acknowledgements

The author would like to thank all colleagues at Novartis and at other pharmaceutical companies for their valuable contributions to the presentation. In particular, the advice and contribution of Jürgen Hubert Blusch and Klaus Vosbeck (Novartis Technical Research & Development, Biotechnology Development) is very much appreciated.

Dr. A. Möritz, Novartis Pharma AG, Technical Research & Development Quality Assurance, WKL-121.211, CH-4002 Basel, Switzerland.
E-mail: Annemarie.moeritz@pharma.novartis.com

Robertson J (ed): PDA/EMEA European Virus Safety Forum.
Dev Biol. Basel, Karger, 2004, vol 118, p 45.

SESSION II

Testing Source Material

Chairs: *P. Korhola* (NAM)
G. Vicari

Robertson J (ed): PDA/EMEA European Virus Safety Forum.
Dev Biol. Basel, Karger, 2004, vol 118, pp 47-53.

Developing New Infectivity Assays

D.N. Galbraith

Formerly with BioReliance, West of Scotland Science Park, Glasgow, U.K.

History of Virology and Virological Techniques

The history of virology as a scientific study is relatively short when compared with other fields of biology such as botany or zoology. The first demonstration of virus infection was by Dmitri Ivanowski in 1892 using "filterable" agents to infect tobacco plants. This initial finding in plants was extended to animals and finally viruses were discovered in humans. Early work demonstrated viruses were capable of causing infectious disease, but the technology to characterise these agents did not exist. Before the Second World War, most virology was performed in laboratory animals, predominately rodents or embryonated chicken eggs. Animals could be used for the isolation and amplification of various viruses, including viruses such as rabies and vaccinia. However, there was a lack of animal models that were able to isolate a number of common human viral diseases such as the Hepatitis viruses. Although limited, the use of animals and eggs in such a way allowed a huge increase in the knowledge base from which virology in more recent years could not have developed. Only with the development of tissue culture techniques were virologists able to further extend our knowledge of these agents. Tissue culture techniques gradually improved since their inception in the early 1900s. The early limitations of this technique were from the risk of contamination by bacteria. Advances in aseptic technique, the use of defined media for growing cells, and the introduction of antibiotics led to scientists discovering and identifying a large number of hitherto unknown viruses. To this extent, it has been suggested that the most significant advance in virology has been the demonstration by John F. Enders that poliovirus could be grown in cell cultures. The recognition that a virus was present in cells can be made by a number of means, the most dramatic observation being cytopathic effect; this causes morphological changes to the cell structures that can frequently lead to cell death. Other observable virus- induced cell changes are inclusion bodies; these are areas of viral proteins or cell structures that can be visualised by cytological staining. Other than morphological changes there are markers on infected cells that can be used to detect virus infection. Haemadsorbtion is a well-established technique where red blood cells can be used to attach onto the haemagglutinin proteins expressed on the surface of some virally infected cells. In a similar way antibodies can specifically identify virus proteins in infected cells where there is no observable cytopathic effect. Recently the advent of molecular amplification techniques has allowed the detection of a number of viruses in cell culture. The application of such cell culture techniques showed for the first time an explanation for the clinical syndromes by an infectious agent and the discovery of a vast array of new viruses.

An Emerging Virus of Concern

Polyomaviruses were once classified as a sub-family of the *Papovaviridae* but now form a separate genus of around 15 species. The viruses are distributed worldwide, typically causing persistent infection in the host animal. Infections occur early in life, at least in the human polyomaviruses studied, and are tissue-specific; frequently observed in the kidney, however, a number of polyomaviruses infect neural cells. Transmission is believed to occur by close contact or by airborne infection but pregnancy appears to reactivate persistent infections in humans and transplacental infection may occur. From the limited studies that are available the host range of polyomaviruses appears to be restricted to the host animal species and near relatives. An interesting feature of infection of the virus in non-competent cells is the malignant transformation by the virus of the cells. The simian virus SV-40 is perhaps the best studied of the polyomaviruses; it does not cause a cytopathic effect when infecting its natural host cells (primate) but will induce tumours in new-born Syrian hamsters. The contamination of batches of human poliovirus vaccines by SV-40 during the early vaccination programmes raised concern that this virus may induce similar pathology in humans [1]. Subsequent studies have not indicated any links between human recipients of the contaminated batches and the subsequent formation of tumours [2].

The first description of *Bovine polyoma virus* (BPyV) in the literature is somewhat difficult to determine. Coakley and colleagues, working in Australia, were the first to describe a *Bovine Polyoma virus* termed WRSV, identified in cells from calf kidney [3]. The virus was identified using immuno-staining techniques and transmission electron microscopy. These studies revealed a virus that was indistinguishable from SV-40. Although this was the first isolation from bovine cells, a similar virus was identified in 1972 and reported by Newman and Smith as a swine papovavirus [4]. The virus was identified by immuno-staining and electron microscopy in porcine cells. As a complicating factor the cells were grown in calf serum. It is possible that the virus reported by Newman and Smith was BPyV that had infected the porcine cells. In a similar way, work carried out by Parry, reported in 1983, initially identified a new polyomavirus in primate cells but this was, on further study, identified as BPyV, probably infecting the primate cells from contaminated calf serum [5]. The accumulation of data through the 1980s indicated that BPyV, unlike a number of other polyomaviruses, was able to infect a variety of cell lines from different species [6]. An important aspect to the natural life cycle of the virus is the lack of pathology observed in cattle infected with the virus. The lack of pathology in cattle can result in infected animals being used as a source of raw materials for the production of biotechnology products. The European regulators have recognised the potential risk from contamination of bovine-derived materials to biotechnology products [7]. Serum from foetal and newborn calves is widely used in biotechnology, mainly to support the growth of cells. Commercially bovine serum is prepared in batches that can be made up from donations of 2,000-4,000 calves. Previous studies have shown that the sero-positivity in cattle is very high, suggesting a widespread level of infection in cattle [8]. Studies on foetal calf serum have subsequently shown that by PCR assays, in excess of 70% of batches were positive for the presence of BpyV-specific sequences [9]. Internal data (Bioreliance Ltd,) from recent batches of foetal calf serum showed over 90% of batches were positive using a specific PCR technique for evidence of viral sequences. Serum from North America, South America, Australia and New Zealand were all shown to be positive [10]. Only a single sample from Europe was found to be negative for BPyV sequences.

Clearly the extent of potential BPyV contamination of bovine derived materials, if proved to be linked to infectious virus, could potentially halt the production of a number of critical human therapies if it became apparent that this virus was a risk. Work by Parry and Gardner has already established the ability of the virus to elicit a humoral immune response in humans occupationally exposed to cattle. There is a need therefore to show that the contamination of bovine-derived materials by BPyV is non-infectious. The need for the development of an infectivity assay was clear.

Three factors are critical to the success of an infectivity assay. These are

1. Appropriate virus stock.
2. Permissive and reliable cell line.
3. Appropriate detection method.

Virus Stock

The initial BPyV virus preparation [obtained from D. Stagg from the Veterinary Laboratory Agency (VLA)] was inoculated on to duplicate 80 cm^2 tissue culture flasks containing Primary Calf Kidney CK cells [obtained from the VLA] and maintained for eight weeks. Media were not supplemented with bovine serum supplements but lamb serum (10%) used in its place [lamb serum was previously checked by BpyV- specific PCR to ensure this was negative (PCR details are contained in [9]). The supernatant was regularly removed from the flasks and stored at or below -70°C. DNA was extracted from aliquots of week one, week four and week eight samples and tested for the presence of BPyV sequences by PCR. The harvests from the tissue culture flasks were pooled and were designated the master virus seed stock. This was characterised and shown to have a genome count of 3.2 x 10^{10} genome equivalents per ml. The whole genome of the virus was confirmed by long-distance PCR and sequencing. Virus was also observed by electron microscopy (Fig. 1).

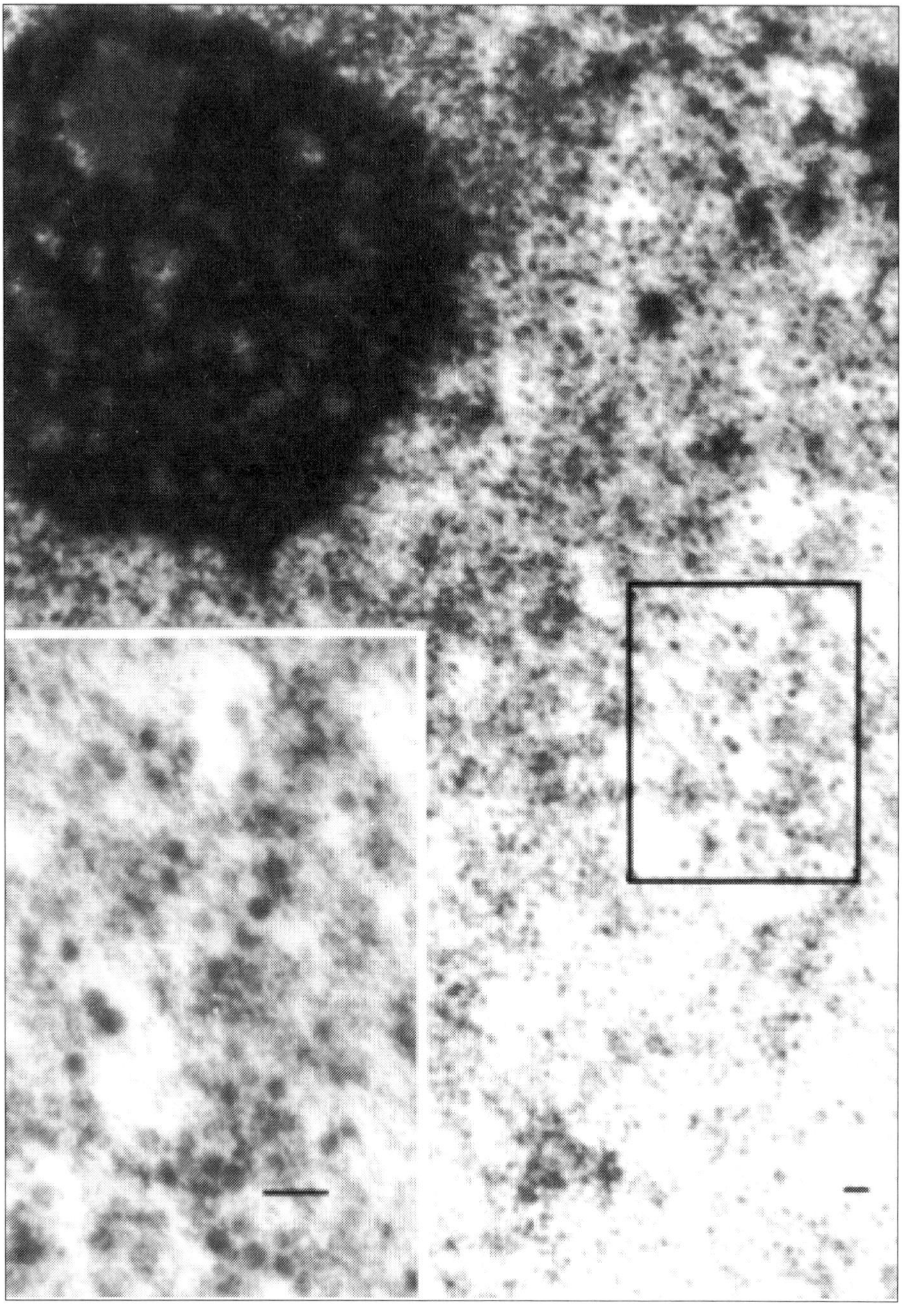

Fig. 1: Electron Micrograph of a nuclear area showing Bovine Polyoma virus in calf kidney cells. Bar ≡ 100nm.

Micrograph courtesy of E.W.Milne, BioReliance (Glasgow)

Inoculation of Bovine, Primate, Human and Rodent Continuous Cell Lines with BPyV

Five different continuous cell lines were inoculated with 10^7g.e. of BPyV and cultured for seven weeks. Negative control flasks were inoculated with media. Supernatants harvested from these cultures and from the equivalent negative detector cells were tested for the presence of BPyV sequences by Q-PCR. There was no evidence of BPyV replication in the human and rodent cell lines. The negative cells remained negative for BPyV sequences throughout the culture period and the levels of BPyV in the positive flasks (CK cells) did not increase significantly.

Optimisation of Growth of BPyV on Calf Kidney Cells

The detection limit of the assay was determined by titration of the virus from approximately 10^7 genome equivalent (g.e.) through 10-fold dilutions to 10^{-1}g.e (g.e. were used as the measure of the amount of virus present as this was the only reliable technique currently available). Harvested supernatants were tested for the presence of BPyV sequences to determine the highest dilution of virus inoculum that gave a positive signal after eight weeks of culture in both replicates.

Optimisation of the concentration of lamb serum was performed by allowing the virus to grow in cells supplemented with different concentrations of serum. Growing the cells at different concentrations influences the growth and passage rates of the detector cells. Lamb serum at reducing concentrations from 10% to 1% was used for the culture of the cells inoculated with virus over an eight-week period. Cells cultured at 2% lamb serum were found to achieve the highest amplification of virus over the eight-week observation period.

During the cultivation of calf kidney cells, it became clear that the inherent nature of these primary cells resulted in extensive variation in cell morphology and on some occasions senescence of all or part of the cells. The variability in cell morphology made distinguishing cells exhibiting cytopathic effect from BPyV from uninfected cells difficult. Therefore an alternative cell line was sought to provide a more suitable detector cell for BPyV infection that was at least as sensitive as calf kidney to the virus. A suitable alternative was Madin Darby Bovine Kidney cells (MDBK, sourced from American Tissue Culture Collection).

Validation of Infectivity Assay Using MDBK Cells

Titration of BPyV on MDBK cells

Duplicate sub-confluent layers of MDBK cells were inoculated with a dilution series of BPyV stock (10^7 to 10^2 genome equivalents [g.e.]). The virus was allowed to adsorb for three hours, then removed and replaced with fresh growth medium. The cultures were maintained with weekly replacement of medium. Supernatants harvested at weeks one, four and eight were tested for the presence of BPyV sequences by QPCR.

Table 1 shows the number of BPyV g.e. recovered from the flasks where the C_T values decreased over time.

Table 1

Inoculum	Timepoint	g.e. recovered / flask
BPyV 10^5	Week 1	8160 g.e.
	Week 4	4.47×10^6 g.e.
	Week 8	6.39×10^9 g.e.
BPyV 10^6	Week 1	5.56×10^5 g.e.
	Week 4	5.7×10^7 g.e.
	Week 8	4.9×10^{10} g.e.
BPyV 10^7	Week 1	8.6×10^6 g.e.
	Week 4	6.04×10^8 g.e.
	Week 8	1.74×10^{11} g.e.

In Table 1 there is a clear increase, over the culture period, in the number of g.e. recovered from both flasks inoculated with 10^7 10^6 10^5 g.e. of BPyV. Reproducibility was observed between the duplicate flasks and the data suggest that replication of the virus has occurred in these cultures. For flasks inoculated with less then 10^5 g.e. of BPyV there was no evidence of replicating virus in the cells or in the flasks inoculated with inactivated BPyV. One unit of infectivity for the qualified BPyV stock was thus defined as 10^5 g.e. BPyV sequences were detected at weeks one, four and eight in those flasks where viral replication was occurring. Naive MDBK detector cells were negative for BPyV sequences. The detection limit (DL) of the assay with this qualified virus stock and the MDBK detector cells was determined to be 10^5 g.e. (the highest dilution of virus inoculum that gave a positive signal after eight weeks of culture in both replicates).

Serum spiking experiments were carried out to determine the presence of inhibitory factors to qualify the test materials. To this end serum was spiked with known titres of BPyV and inoculated on to MDBK monolayers as described above. BPyV was spiked at 10^5 to 10^7 genome equivalent of virus. In all spikes virus was detected after passage on MDBK cells. This indicates there was no inhibitory effect on the detection of BPyV by the presence of foetal calf serum.

Conclusion

It has thus been demonstrated that replicating BPyV can be detected using an in vitro culture system coupled with Q-PCR. Cytopathic effect was not obvious in the CK and MDBK cell cultures and could not be used as an end-point to the assay, hence the reason for the PCR, although an immuno-fluoresence (IF) assay may be developed in the future. Quantitative PCR is important to be used in this assay as there can be an extremely high level of BPyV genomes present in foetal and newborn calf serum. A quantitative increase in the level of virus is necessary to show replicating virus is present to distinguish this from background. Switching the detector cells from CK to MDBK cells provides a more reliable method of amplification of the virus with no loss of sensitivity. Replication of virus appeared to be detectable earlier in the continuous cell line rather than the primary. This may be due to the cell cycle of the continuous cells being in a phase more able to initiate replication. The long culture period for detection of replicating virus appeared necessary irrespective of the cell substrate. This mimics the replication of SV-40 virus where high multiplicity of infection leads to more rapid virus replication due to the necessity to achieve critical levels of virus early proteins. There also appears to be a high level of defective virus present in the culture systems (10^5 genome equivalents to one infectious unit). This is known for SV-40 also where deletions, inversions etc. are common but survive as long as "wild-type" virus remains [11].

A recently finalised European guideline has suggested that a suitable assay for the detection of BPyV should be performed on bovine serum used for the manufacture of human biological products [7]. Data already exist that the viral genome can be detected widely in bovine materials and that the PCR-based assay, which is widely available, is likely to be of little value. Therefore the infectivity assay, described here, provides an alternative method to identify those materials replicating virus, where there is immediate concern. In the future it may be possible to show that using some common inactivation techniques, this virus may be eliminated as a risk in products.

References

1 Ferber D: Public health. Creeping consensus on SV40 and polio vaccine. Science 2002;25;298(5594):725-727.

2 Strickler HD, Rosenberg PS, Devesa SS, Hertel J, Fraumeni JF Jr, Goedert JJ: Contamination of poliovirus vaccines with simian virus 40 (1955-1963) and subsequent cancer rates. Journal of the American Medical Association 1998;279(4):292-295.

3 Coakley W, Maker D, Smith VW: A possible bovine polyomavirus. Archives of Virology 1980;66:161-166.

4 Newman JT, Smith KO: Characteristics of a Swine Papovavirus. Infection and Immunity 1972;5(6):961-967.

5 Parry JV, Lucas MH, Richmond JE. Gardner SD: Evidence for a bovine origin of the polyomavirus detected in foetal rhesus monkey kidney cells, FRhK-4 and -6. Archives of Virology 1983;78:151-165.

6 Schuurman R, Van Steenis B, Van Strien A, Van Der Noordaa J, Sol C: Frequent detection of bovine polyomavirus in commercial batches of calf serum by using the polymerase chain reaction. Journal of General Virology 1991;72:2739-2745.

7 The European Agency for the Evaluation of Medicinal Products, Evaluation for Human Use, Note for Guidance on the Use of Bovine Serum in the Manufacture of Human Biological Medicinal Products (CPMP/BWP/1793/02), 18 June 2003.

8 Parry JV, Gardner SD: Human exposure to bovine polyomavirus: A zoonosis? Archives of Virology 1986;87:287-296.

9 Kappeler A, Lutz-Wallace C, Sapp T, Sidhu M: Detection of Bovine polyomavirus contamination in fetal bovine sera and modified live viral vaccines using polymerase chain reaction. Biologicals 1996;24:131-135.

10 Nairn C, Lovatt A, Galbraith DN: Detection of infectious bovine polyomavirus. Biologicals 2003;31(4):303-306.

11 Lee TNH, Brockman WW, Nathans D: Evolutionary variants of SV-40: Cloned substituted variants containing multiple initiation sites for DNA replication. Virology 1975;66:53-69.

Dr. D.N. Galbraith, MedImmune Ltd., Boulevard Industrial Estate, Speke, Liverpool, L24 9JW, U.K.
E-mail: galbraithd@medimmune.com

Robertson J (ed): PDA/EMEA European Virus Safety Forum.
Dev Biol. Basel, Karger, 2004, vol 118, pp 55-63.

Qualification of Virus/Cell Systems for their Intended Use

A. Gröner

Aventis Behring, Virology, Marburg, Germany

INTRODUCTION

In the testing and evaluation of the virus safety of biological products derived from well characterised cell lines of human or animal origin, or from blood, urine or other biological fluids of human or animal origin, cell lines are used to detect viruses. These viruses may be contaminants of the starting material or of raw materials used as media components in the production of defined biological products. In virus validation studies, the viruses to be detected are deliberately added to the starting material or their intermediates to evaluate the capacity of the manufacturing process to remove and/or inactivate the viruses.

The suitability of cell lines for the isolation and/or detection of viruses, and the interpretation of results, will be discussed in this paper, taking into consideration the appropriate guidelines such as ICH Topic Q 5 A (Europe: CPMP/ICH/295/95 or USA: Docket No. 96D-0058) and CPMP/BWP/268/95 as well as CPMP/BWP/269/95 – rev.3.

VIRUS SAFETY EVALUATION OF BIOTECHNOLOGY PRODUCTS

As the risk of virus contamination is a common feature of all biotechnology products derived from cell lines, the safety of these products with regard to potential transmission of viruses can only be reasonably assured by the application of a stringent virus testing programme along with virus validation studies.

Source of virus contaminations

Virus contamination may arise from the original source of a cell line used for production or from adventitious introduction of viruses during the production processes. Therefore, the Master Cell Bank (MCB) has to be tested for the presence of viruses which can be introduced into the MCB due to (i) contamination of animals or humans from which the cell line is derived; (ii) contamination of reagents used during the cultivation of cell lines, such as animal serum components; (iii) contamination during cell handling, and (iv) use of a virus to establish a permanent cell line. Viruses may be detected by methods not involving cell cultures such as antibody production tests (for the detection of known/conceivable viruses that replicate in a specific animal test system or if present in large amounts in the test sample), in vivo virus screening (detection of a broad range of viruses if they cause symptoms/disease in the test animals/system), transmission electron microscopy (the physical detection of viruses and virus-like particles; [1]), reverse transcriptase assays (RT; for the detection of retroviruses; [2]) and NAT (or PCR) (for the detection of the nucleic acid of known/conceivable viruses). In vitro virus screening using cell lines will detect, as can in vivo assays, a broad range of viruses if diagnosis of the infected cell culture is feasible.

Contamination of biotechnology products that is due to adventitious viruses may occur during the production process by several routes such as the use of contaminated (i) biological reagents such as cell culture medium components; (ii) in-process material such as monoclonal antibody affinity columns; (iii) excipients used for formulation; (iv) contamination during cell and medium handling and (v) use of a virus for the induction of expression of a specific gene(s) encoding a desired protein. Therefore, in addition to Master Cell Banks (MCB) as well as Working Cell Banks (WCB) derived from the MCB, cells at the limit of in vitro cell age used for production should be tested (this testing regime is described in the ICH guideline Topic Q5A). To detect contaminating viruses with a high probability, cells, cell lysate, and culture medium should be tested.

The detection of potential viral contaminants using indicator cell cultures requires a range of suitable cell lines to cover all potentially contaminated materials used during production. For practical reasons, the number of indicator cell lines should be as low as possible; but should consist of a range of cell lines for the purpose of detecting all viruses that could conceivably be present in the materials to be tested. As not all viruses cause a cytopathic effect (CPE) in an indicator cell line, further methods to demonstrate the presence of a virus in an indicator cell line have to be used, such as haemaglutination/haemadsorption using erythrocytes of different species (e.g., chicken, guinea pig, Rhesus monkey or human type O) or, for specific potential contaminants, immunofluorescence or nucleic acid amplification techniques (NAT), testing may be an option. For testing the MCB, cell lines from the same or a closely related species should be used primarily to detect viruses originating from a potentially infected source animal (or human); by using indicator cell lines from the same species from which the production cell line was established, a potential species barrier may be avoided. Viruses originating from medium components should be tested using bovine cell lines (for bovine serum derived viruses) or porcine cell lines for trypsin-derived viruses. Each MBC/WCB should be tested with appropriate cell lines able to detect virus contaminants derived from medium components used in MBC/WCB passaging or from any cell line that may have been in contact with the MCB/WCB during the whole history of the production cell line. Furthermore, virus contamination that may arise during cell and medium handling should be tested using, in addition, human cell lines.

The unprocessed bulk, the (pooled) harvest of cells and culture media, constitutes one of the most suitable intermediates for the detection of adventitious virus contamination. Therefore, a mixture consisting of both intact and lysed cells and their cell culture supernatants should be tested. Besides, monitoring cell culture parameters and deviations from standard production parameters such as total oxygen consumption, amount of buffering substances, and cell culture medium (concentrate) used, may indicate potential adventitious virus contamination of a defined fermentation batch.

Virus detection

Virus contamination of cell cultures used for the production of biotechnology products that results in a CPE in the cell line [3] is easily detected and, therefore, not the scope of this paper. In addition, endogenous viruses (such as retroviruses) are also not discussed here because these viruses are general contaminants of cells; nevertheless, the presence of infectious or non-infectious retroviruses in the cell line and the unprocessed bulk has to be documented [2]. In virus validation studies, the reduction of these viruses by the manufacturing process of biotechnology products has to be demonstrated. As cell culture process changes may have an impact on the expression level of endogenous retrovirus expression [4], the virus reduction capacity has to be assessed after each process change.

As cell cultures may harbour viruses other than endogenous retroviruses, these contaminants, deriving most probably from raw material (when the MCB has been thoroughly tested), have to be tested for and identified. A stringent approach to detect potential contaminants, especially when they do not interfere significantly with cell replication, has to be implemented [5].

Viruses that originate from an infected animal (or human) are well known contaminants of cell cultures, e.g., primary Rhesus kidney cell cultures were contaminated with SV40 [6] or tumour cell lines infected with LCMV [7]. In addition, it has to be considered that all (especially rodent) cell lines express endogenous retroviruses [1,2,8,9].

Contaminants of material used during the replication of the cells have to be considered starting from the establishment of the cell line up to the production of the biotechnology product. Therefore, cell culture medium and their ingredients such as foetal bovine serum as well as trypsin used during passaging have to be assessed as potential sources of virus contamination.

A papovavirus was isolated from Cynomolgus monkey kidney cell cultures after a prolonged period of culturing, due to the appearance of a cytopathic effect, but was non-cytopathic in BSC-1, Vero, human embryonic and calf kidney cells. This papovavirus was originally thought to be of simian origin, having been isolated from cell lines derived from the stump-tailed macaque [10] and also from Rhesus monkeys, as from foetal Rhesus monkey kidney cell lines FRhK-4 and FRhK-6 [11]. As a high proportion of cattle, but no monkeys, were positive for antibodies to the virus, the virus was re-classified as bovine polyomavirus (BPyV). BPyV is a contaminant of foetal bovine serum easily detectable by NAT; however, as NAT does not distinguish between replicative and non-replicative virus, a cell culture assay should be used for the detection of BPyV in serum [12].

In addition to polyomaviruses, parvoviruses can be detected in cell cultures due to contamination of serum, trypsin and other medium components [5,13-17].

It is also well established knowledge that a high percentage of batches of foetal bovine serum carry bovine viral diarrhoea virus (BVDV), resulting in a wide range of cell lines being contaminated by this virus. A survey of cell lines originating from a wide range of animal species, which were obtained from the American Type Culture Collection, revealed BVDV antigen or BVDV RNA in 13 of 41 cell lines: cell lines from cattle, deer, sheep, bison, rabbit, cat, swine and monkey were reactive for BVDV or could be deliberately infected by BVDV whereas cell lines from hamster, dog, human and mosquito remained non-reactive [18]. As the detection of BVDV (and other pestiviruses) is hampered because most pestivirus strains are non-cytopathic in cell cultures and immunological assays may be insensitive for the detection of low-titre contamination, appropriate NAT assays should be used for the exclusion of perceived BVDV contamination. However, it has to be cautioned that foetal bovine serum may contain BVDV genomic sequences, but no infectious BVDV, possibly resulting in a false-positive interpretation for BVDV viral contamination.

As cell lines infected with a non-cytopathic BVDV isolate and subsequently superinfected with a cytopathic BVDV isolate were refractive to the cytopathic effects of the cytopathic BVDV isolate, this interference study could be used to detect non-cytopathic BVDV in a cell culture or in an indicator cell line, after some passages of the culture to facilitate the replication of the non-cytopathic BVDV isolate in most of the cells [19].

Foetal bovine serum and trypsin should be thoroughly screened for adventitious agents before a defined batch of these components is used for culturing cells; this screening should include cell replication properties, NAT and infectivity testing for adventitious viruses such as BVDV, BPyV and parvoviruses, as well as, for certain requirements, testing for antibodies against defined viruses, if the cell culture is to be used for the detection of these viruses. If inactivated serum and trypsin is used [20], NAT may give an inappropriate result because viral genome from non-infectious viruses may be detected. However, despite thorough screening, such medium components may result in the infection of a cell culture due to the fact that for all screening assays, the assay sensitivity and sampling of the material to be tested has to be considered. As such, screening assays will achieve a "non-reactive" but never a "negative" status, and continuous testing of cell cultures over time/passage is recommended.

Infectivity assays can be considered the most appropriate detection method for infectious viruses in cell cultures or in cell culture medium if appropriate cell lines with a high sensitivity for a range of viruses are used. However, this high sensitivity should be controlled routinely, as we detected in our own studies a contaminating enveloped virus in a PH-2/F-MC cell culture, which resulted in interference with the detection of a range of certain viruses and significantly lower titres for other viruses, as shown in Table 1. Therefore, an adventitious virus or a low titre of an adventitious virus can remain undetected without appropriate controls if a persistently infected cell line is used as an indicator cell line. As shown in Table 2, the adventitious agent replicated very slowly in the cell culture. When poliomyelitis virus type 1 was used as a model virus to detect the presence of the interfering adventitious virus, no significant decrease in sensitivity of the cell line to polio infection could be demonstrated for more than a month, but within two months the cell line was non-susceptible to polio virus infection. The persistently infected cell clone resembled uninfected cells in most respects as cell morphology (microscopic examination) and doubling time.

Table 1: Susceptibility of a persistently infected cell clone (PH-2/F-MC) in comparison to the non-infected cell line (PH-2), to a panel of viruses.

Virus inoculum	**Virus titre [log10] in cell line**		**Difference in virus titre [log10]**
	PH-2	PH-2/F-MC	
Poliomyelitis virus -1	8.7	≤ 1.5	≥ 7.2
HHV-1	6.9	≤ 1.5	≥ 5.4
MeV	6.1	≤ 1.5	≥ 4.6
SINV	6.0	1.9	4.1
SV40	4.7	2.5	2.2

Table 2: Susceptibility of PH-2 to poliovirus infection after inoculation of PH-2/F-MC cell culture supernate. The PH-2 cell culture was passaged 1:6 once per week and its susceptibility to poliomyelitis virus 1 tested at different passages.

	Polio virus titre [log10 CCID50/ml]				
PH-2 inoculated with supernatant of PH-2 / F-MC	**passage 1**	**passage 3**	**passage 5**	**passage 7**	**passage 8**
	5.4	5.2	4.7	≤ 2.5	≤ 0.5

For the detection of a virus in the MCB/WCB, that has been used to establish a permanent cell line, or for the detection of a virus in the unprocessed bulk of the biotechnology product, that has been used to induce the expression of a desired protein, a cell line proved to be susceptible to the virus should be used.

Virus/Cell Systems Used in Virus Validation Studies

In virus validation studies, known viruses are spiked into product intermediates to evaluate the capacity of a manufacturing step – starting with that intermediate – to inactivate and/or remove the added virus(es). Therefore, cell lines known to be susceptible to the virus(es) in the study should be used. Nevertheless, the sensitivity of the cell line for virus replication should be demonstrated using a standard virus sample as an appropriate control, whose titre should correspond to the established titre for that virus stock. If the susceptibility of the indicator cell line is diminished (e.g., by a persistent virus infection as in PH-2/F-MC, Table 1), a low virus titre in the final sample of the process step assessed may result in the overestimation of the capacity of the manufacturing step to reduce viruses.

In contrast, as adeno-associated viruses (AAV) can persistently infect cell lines, there is a possibility that the inoculation of such a cell line with replication-defective herpesviruses may result in a cytopathic effect due to the helper function of herpesvirus replication genes [21-23]. Therefore, the virus reduction capacity of a manufacturing step may be underestimated when the virus titre in the sample to be tested is based solely on CPE in the inoculated cells.

As discussed by Willkommen [24], assessing the pasteurisation step of albumin – manufactured according to the European Pharmacopoeia monograph – resulted in considerably different HAV inactivation factors when studied by different laboratories (Table 3). Although no details of the HAV infectivity detection systems were

Table 3: Inactivation of HAV by the pasteurisation step of human serum albumin production (4 or 5% protein) manufactured according to Ph Eur Monograph [presented by H. Willkommen, Regulatory Issues Surrounding Blood Product Safety at IIR 3rd International Conference "Optimising Blood Product Safety – London 2000"].

Product/Manufacturer	**A***	**B**	**C**	**D**	**E**	**F**
HAV reduction factor [log10]	≥ 6.6	2.9	3.0	5.7	3.8	4.5-5.8

* Product/manufacturer A is determined to be Aventis Behring, based on data submitted to regulatory authorities.

disclosed, it can be assumed that several laboratories were using the FRhK-4 cell line for virus titration. According to our own experience and other information (G. Pauli/H.R. Gelderblom, pers. comm.), this cell line is contaminated with a paramyxovirus-like virus. We developed in our laboratory, an FRhK-4 derivative called RhK that is free from this virus. Using this cell line, we were able to demonstrate complete inactivation of HAV by pasteurisation of albumin (Product/Manufacturer A in Table 3, concluded due to data submitted to regulatory authorities). Whether the discrepancy in the results in HAV inactivation factors are due to the contamination of the cell line, however, has to be evaluated in further studies.

Besides viruses, other adventitious agents may be present in cell cultures if appropriate controls are not implemented to detect contamination and discard any contaminated cell stock. For instance, mycoplasma infections can alter the cytopathic effect of a virus infected cell line [25].

Besides contamination by adventitious viruses/pathogens, the susceptibility of an indicator cell line may be altered due to cross-contamination by other cell lines [26-28]. Again, a thorough assessment of the susceptibility of a given indicator cell line by a range of pre-defined viruses may reveal this contamination.

Conclusion

To avoid biologicals that are contaminated with viruses being administered to humans, as has happened in the past (Table 4), a thorough screening of the starting, as well as in-process material, including excipients and stabilisers (e.g., Yellow Fever vaccine contaminated with HBV present in human serum used as stabiliser [36]) is essential. To detect these adventitious agents by an in vitro screening approach, a range of appropriate cell lines with sufficient susceptibility and sensitivity has to be used. The cell lines have to be routinely assessed for their intended use, i.e., sensitivity to a predefined panel of viruses. Such a thorough screening of biotechnology products with regard to cell culture, medium components, unprocessed bulk and, furthermore, the (downstream) manufacturing process with its capacity to inactivate and/or remove viruses, documented in virus validation studies, will result in a medicinal product of adequate safety.

Table 4: Contaminations of cells used in the production of biotechnology products.

Virus	Probable source	Material tested	Detection in	Clinical consequence	Product	Reference
BVDV	FBS	Final product	Vaccine recipients	Probable virus transmission	Animal vaccine	[29]
Bluetongue virus	Unknown	Final product	Vaccine recipients	Probable virus transmission	Animal vaccine	[30]
BVDV	FBS	Final product	Final product	None ?	Animal vaccines	[31]
BPyV	FBS	Final product and FBS	Final product and FBS	None ?	Animal vaccine	[32]
SV40	Monkey (primary kidney cells)	Final product	Vaccine recipients	Probable virus transmission	Human vaccine	[33]
EHDV	FBS	Cell culture	Cell culture (CPE in production cell culture)	None (product not released)	Rec-protein	[3]
BVDV	FBS	Final product	Final product	None ?	Rec-protein	[34]
MVM	Medium components	Unprocessed bulk	Unprocessed bulk	None (product not released)	Rec-protein	[35]

References

1 Reid GG, Milne EW, Coggins LW, Wilson NJ, Smith KT, Shepherd AJ: Comparison of electron microscopic techniques for enumeration of endogenous retrovirus in mouse and Chinese hamster cell lines used for production of biologics. J Virol Meth 2003;108:91-96.

2 Shepherd AJ, Wilson NJ, Smith KT: Characterisation of endogenous retrovirus in rodent cell lines used for production of biologicals. Biologicals 2003;31:251-260.

3 Rabenau H, Ohlinger V, Anderson J, Selb B, Cinatl J, Wolf W, Frost J, Mellor P, Doerr HW: Contamination of genetically engineered CHO-cells by epizootic haemorrhagic disease virus (EHDV). Biologicals 1993,21:207-214.

4 Brorson K, De Wit C, Hamilton E, Mustafa M, Swann PG, Kiss R, Taticek R, Polastri G, Stein KE, Xu Y: Impact of cell culture process changes on endogenous retrovirus expression. Biotechnol Bioeng 2002;80:257-267.

5 Garnick RL: Raw materials as a source of contamination in large-scale cell culture, in Brown F, Griffiths E, Horaud F, Petricciani JC (eds): Safety of Biological Products Prepared from Mammalian Cell Culture. Dev Biol Stand. Basel, Karger, 1998, vol 93, pp 21-29.

6 Shah K, Nathanson N: Human exposure to SV40: Review and comment. Am J Epidemiol 1976;103:1-12.

7 Mahy WJ, Dykewicz C, Fisher-Hoch S, Ostroff S, Tipple M, Sanches A: Virus zoonoses and their potential for contamination of cell cultures, Brown F, Horaud F (eds): Virological Aspects of the Safety of Biological Products. Dev Biol Stand. Basel, Karger, 1991, vol 75, pp 183-189.

8 Weiss RA: Perspectives on endogenous retroviruses in normal and pathological growth. Curr Top Microbiol Immunol 1982;98:4127-4132.

9 Dinowitz M, Lie YS, Low MA, Lazar R, Fauth C, Potts B, Sernatinger J, Anderson K: Recent studies on retrovirus-like particles in Chinese hamster ovary cells, in Brown F, Esber EE, Williams M (eds): Continuous Cell Lines – An International Workshop on Current Issues. Dev Biol Stand. Basel, Karger, 1992, vol 76, pp 201-207.

10 Wognum AW, Sol CJ, van der Noordaa J, van Steenis G, Osterhaus AD: Isolation and characterisation of a papovavirus from cynomolgus macaque kidney cells. Virology 1984;134:254-257.

11 Parry JV, Lucas MH, Richmon JE, Gardner SD: Evidence for a bovine origin of the polyomavirus detected in foetal rhesus monkey kidney cells, FRhK-4 and -6. Arch Virol 1983;78:151-165.

12 Nairn C, Lovatt A, Galbraith DN: Detection of infectious bovine polyomavirus. Biologicals 2003;31:303-306.

13 Kilham L, Olivier LJ: A latent virus of rats isolated in tissue culture. Virology 1959;7:428-437.

14 Wozniak J, Hetrick F: Persistent infection of a rat nephroma cell line with Kilham rat virus. J Virol 1969;4:313.

15 Hallauer C, Kronauer G, Siegl G: Parvoviruses as contaminants of permanent human cell lines. I. Virus isolations from 1960 – 1970. Arch Gesamte Virusforsch 1971;35:80-90.

16 Croghan DL, Matchett A, Koski TA: Isolation of porcine parvovirus from commercial trypsin. Appl Microbiol 1973;26:431-433.

17 Nettleton PF, Rweyemamu MM: The association of calf serum with the contamination of BHK 21 clone 13 suspension cells by a parvovirus serologically related to the minute virus of mice. Arch Virol 1980;64:359-374.

18 Bolin SR, Ridpath JF, Black J, Macy M, Roblin R: Survey of cell lines in the American Type Culture Collection for bovine viral diarrhoea virus. J Virol Meth 1994;48:211-221.

19 Moennig V, Plagemann PG: The pestiviruses. Adv Virus Res 1992;41:53-98.

20 Erickson GA, Landgraf JG, Wessman SJ, Koski TA, Moss LM: Detection and elimination of adventitious agents in continuous cell lines, in Hayflick L, Hennessen W (eds): Continuous Cell Lines as Substrates for Biologicals. Dev Biol Stand. Basel, Karger, 1989, vol 70, pp 59-66.

21 Handa H, Carter BJ: Adeno-associated virus DNA replication complexes in herpes simplex virus or adenovirus-infected cells. J Biol Chem 1979;254:6603-6610.

22 McPherson RA, Rosenthal LJ, Rose JA: Human cytomegalovirus completely helps adeno-associated virus replication. Virology 1985;147:217-222.

23 Stracker TH, Cassell GD, Ward P, Loo YM, van Breukelen B, Carrington-Lawrence, SD, Hamatake, RK, van der Vliet PC, Weller SK, Melendy T, Weitzmann MD: The Rep protein of adeno-associated virus type 2 interacts with single-stranded DNA-binding proteins that enhance viral replication. J Virol 2004;78:441-453.

24 Willkommen H: Regulatory Issues Surrounding Blood Product Safety – IIR 3rd International Conference "Optimising Blood Product Safety – London 2000".

25 Lemaître M, Henin Y, Destouesse R, Ferrieux C, Montagnier L, Blanchard A: Role of mycoplasma infection in the cytopathic effect induced by human immunodeficiency virus type 1 in infected cell lines. Infect Immun 1992;60:742-748.

26 Nelson-Rees WA, Daniels DW, Flandermeyer RR: Cross-contamination of cells in culture. Science 1981;212:446-452.

27 Orua H, Fujii R, Hamano, M, Kuzuya, M, Nakajima H, Ohata R, Mori T: Detection of HeLa cell contamination – presence of human papilloma virus 18 DNA as HeLa markers in JTC-3, OG and OE cell lines. Jpn J Med Sci Biol 1997;50:161-167.

28 Drexler HG, Dirks, WG, MacLeod RAF: False human hematopoietic cell lines: cross –contaminations and misinterpretations. Leukemia 1999;13:1601-1607.

29 Kreeft HA, Greiser-Wilke I, Moenning V, Horzinek MC : Attempts to characterize bovine viral diarrhea virus isolated from cattle after immunization with a contaminated vaccine. Dtsch Tierärztl Wochenschr 1990;97:63-65.

30 Wilbur LA, Evermann JF, Levings RL, Stoll IR, Starling DE, Spiller CA, Gustafson GA, McKernan AJ : Abortion and death in pregnant bitches associated with a canine vaccine contaminated with bluetongue virus. J Am Vet Med Assoc 1994;204:1762-1765.

31 Harasawa R: Adventitious pestivirus RNA in live virus vaccines against bovine and swine diseases. Vaccine 1995;13:11-103.

32 Kappeler A, Lutz-Wallace C, Sapp T, Sidhu M: Detection of bovine polyomavirus contamination on fetal bovine sera and modified live viral vaccines using polymerase chain reaction. Biologicals 1996;24:131-135.

33 Hilleman MR: Discovery of simian virus 40 (SV40) and its relationship to poliomyelitis virus vaccines, in Brown F, Lewis AM Jr (eds): Simian Virus 40 (SV40): A Possible Human Polyomavirus. Dev Biol Stand. Basel, Karger, 1998, vol 94, pp 183-190.

34 Harasawa R, Sasaki T: Sequence analysis of the 5' untranslated region of pestivirus RNA demonstrated in interferons for human use. Biologicals 1995;23:263-269.

35 Garnick RL: Experience with viral contamination in cell culture, in Brown F, Lubiniecki AS (eds): Viral Safety and Evaluation of Viral Clearance from Biopharmaceutical Products. Dev Biol Stand. Basel, Karger, 1996, vol 88, pp 49-56.

36 Findlay GM, MacCallum FO: Note on acute hepatitis and yellow fever immunization. Trans R Soc Trop Med Hyg 1937;31:297-308.

Dr. A. Gröner, Virology, ZLB Behring GmbH, Emil-von-Behring Strasse 76, D-35041 Marburg, Germany. E-mail: Albrecht.Groener@ZLBBehring.com.

Robertson J (ed): PDA/EMEA European Virus Safety Forum.
Dev Biol. Basel, Karger, 2004, vol 118, pp 65-70.

West Nile Virus NAT Testing

J. Saldanha, C. Cameron

Canadian Blood Services, Ottawa, ON, Canada

INTRODUCTION

West Nile Virus (WNV) is a mosquito-borne flavivirus, which is endemic in Africa, Asia and Europe [1,2]. Birds are the primary vertebrate host for WNV, while humans and other mammals are incidental hosts. Most WNV infections in humans are asymptomatic; only 20% develop a mild febrile illness and less than 1% develop encephalitis or meningitis. Transmission of WNV by blood transfusion is possible given that infection, including sub-clinical infection, is associated with a transient viraemia [3,4]. However, transfusion-associated transmission of WNV or related Japanese encephalitis serogroup viruses was not documented before 2002.

The virus was first detected in the U.S.A. in 1999 in Queens, New York [5,6], since when it has spread rapidly across the country. This pattern of spread is most probably due to long-distance transport of WNV in infected migratory birds combined with local dissemination by mosquitoes. During the 2002 WNV epidemic in the U.S.A., there were over 4000 cases of WNV (Centers for Disease Control and Prevention, CDC, unpublished data). A total of 23 persons were reported to have acquired WNV after receipt of blood components from 16 WNV viraemic donors [7-9]. In Canada, five transfusion-transmitted WNV cases were reported (personal communication). Two of these cases were linked to the same WNV viraemic donor, in one case no donor was implicated and in the remaining two cases no investigation was conducted due to the large number of transfused components. Transmission of WNV from an organ donor to four transplant recipients in the U.S.A. was also documented in 2002 [10]. At a meeting held in Washington DC, U.S.A., in November 2002, the FDA urged kit manufacturers to develop a NAT assay for testing donors for WNV RNA by the following summer, when the expected peak of the WNV epidemic was predicted to occur. By the end of August 2003, forty-four states in the U.S.A. (including Washington DC) were endemic for WNV and over 3000 cases of WNV were recorded [11].

WNV Testing in the U.S.A. and Canada

Commercial assays

Commercial WNV NAT assays for donor screening have been developed by two kit manufacturers, Roche Molecular Systems and Chiron. The Roche TaqScreen WNV test is used by approximately 20% of U.S.A. blood banks and by both Canadian blood operators (Canadian Blood Services, CBS, and Hema Quebec, HQ). Samples are tested in pools of six or individually. The Chiron Procleix assay is used by 80% of the U.S.A. blood banks and samples are tested in pools of 16 or individually.

CBS in-house WNV NAT assay

In April 2003, CBS made the decision to establish an in house assay for WNV NAT testing of blood donations in the event that a commercial test was not available by the start of the WNV season. The in-house assay consisted of manual extraction of a 200µl plasma sample using the Qiagen mini blood DNA kit followed by RT/PCR on 10µl of the extracted RNA (equivalent to 33.3µl plasma sample) using the Qiagen RT/PCR probe kit and the Stratagene MX4000 real-time instrument. The RT/PCR kit contains ROX dye, which is used to normalise the results from each well. Preliminary evaluation using a heat-inactivated WNV sample from the Health Canada National Microbiology Laboratory, Winnipeg, Canada showed that the assay had a 95% detection limit of 95 copies/ml (based on a titre of 2 x 10^9 copies/ml for the Health Canada sample). The in-house assay used primers from the 3' non-coding region of the NY99 strain [12]. The primers will detect WNV strains from Romania, Egypt, Italy, Kenya and Kunjin but not dengue virus type 2, yellow fever virus, SLE virus and JE virus. Samples were tested in pools of eight donations. The pools were prepared on a Qiagen BioRobot 8000 followed by manual extraction and RT/PCR on the MX4000 as described. The primers amplified a 10^3 base pair region, and a FAM-labelled TaqMan probe was used for detection of the amplicons. For routine testing, an internal control, consisting of Armoured RNA (Ambion) containing 242 base pairs from the WNV NS5-2 region, was extracted and co-amplified. A TaqMan probe labelled with HEX was used for detection of the internal control amplicons. The internal control was added to a final concentration of 3000 copies/ml to the samples before extraction. A positive control containing approximately 2000 copies of heat-inactivated WNV was run in duplicate together with duplicate negative controls (WNV-negative, pooled human plasma).

Testing of Canadian donations using the in-house assay started on 17th June 2003. A maximum of 576 samples were tested per day and five runs were completed by 23rd June 2004 when screening with the Roche WNV assay was implemented. The samples tested were from Toronto, Ottawa, Hamilton and London and included all apheresis platelet donations, all donations for paediatric use and approximately 62% of random donor platelets.

Results of the in-house assay

Five out of the six assay runs were successful (the positive and negative controls passed and the internal control was detected in the majority of samples). Discrepant results included three failed internal controls, on three different runs, which were resolved by re-extracting the samples and amplifying both the original RNA extract and the re-extracted RNA. A single pool sample gave an initial positive result. However, re-extraction of the pool and the related eight individual donations, and amplification of both the original pool RNA and the re-extracted RNA samples, showed that this was a false-positive result. One complete run of 576 samples failed, as three of the four positive controls were not detected. This was subsequently shown to be a failure at the RT/PCR step as re-amplification of the original RNA samples gave the expected results for the controls. However, due to time constraints, these samples were released as 'Not tested'. A total of 3148 samples were tested using the in-house assay and 2813 samples (excluding the samples in the failed run) were released as non-reactive for WNV.

Roche TaqScreen testing

All donor samples were tested using the Roche assay at two centres, Toronto (from 23rd June 2003) and Calgary (from 2nd July 2003). Of 190,276 samples tested (up to 15th September 2003), 14 were positive for WNV. Eleven of these samples were tested in pools of six donations and one sample was tested individually. Two samples were identified when single donor testing was done in WNV high prevalence areas. Plasma units from all the positive donations were retested using the in-house assay as a supplemental NAT assay. Twelve of the fourteen samples were also found to be positive in the in-house assay. Apart from one sample with a titre below the 95% detection limit, the titres of the samples ranged from approximately 2×10^3 to approximately 7×10^4 copies/ml (Fig. 1). Positive donations were first detected in mid July, peaked in mid-August and the last positive donation was identified at the beginning of September (Fig. 2).

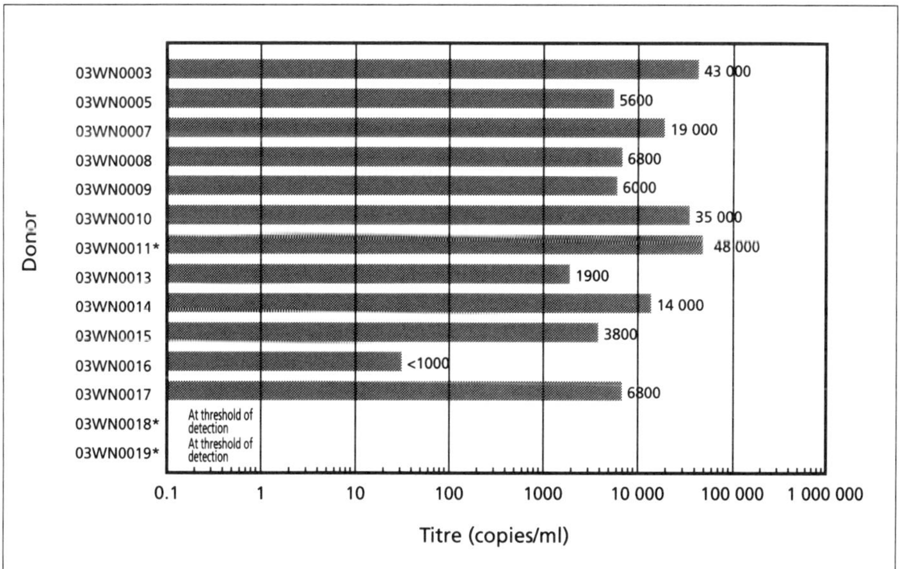

Fig. 1 : Titres of positive donations (in-house assay).

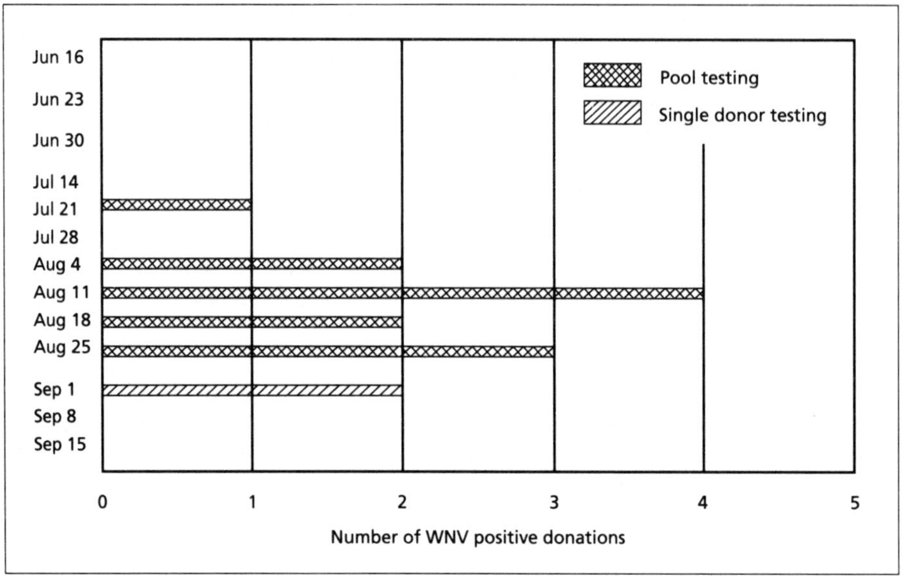

Fig. 2 : Incidence of WV-positive donations.

Single donor testing

Single-donor testing was introduced in the U.S.A. as a limited research study to determine the sensitivity of single donor versus mini-pool testing. Some blood operators initiated testing of single donors in high prevalence areas. In Canada, testing of single donations for a limited number of donations was initiated in Saskatchewan at the beginning of September. This province had the highest WNV prevalence in Canada (Tables 1 and 2). Compared with 2002, where the highest

Table 1 : WNV prevalence in Canada.

Province	Human cases	Blood donors	TTI
British Columbia	2	0	0
Alberta	146*(2)*	3	0
Saskatchewan	219	10	0
Manitoba	10	1	0
Ontario	25*(395)*	0*(1)*	0*(5)*
Quebec	1*(19)*		
Nova Scotia	1	0	0
Total	404*(416)*	14*(1)*	0*(5)*

Cases to 15th September 2003
TTI: transfusion-transmitted infections
( ) Cases in 2002

Table 2: WNV human cases in Canada.

Province	Human cases (per 10^6)
British Columbia	0.51*(0.00)*
Alberta	49.08 *(0.69)*
Saskatchewan	223.71*(0.00)*
Manitoba	8.93 *(0.00)*
Ontario	2.19 *(34.62)*
Quebec	0.14 *(2.63)*
Nova Scotia	1.10 *(0.00)*
Total	13.46 *(13.86)*

Cases to 15th September 2003
( ) Cases in 2002

prevalence was in Ontario, the peak of WNV infection had moved westwards in 2003 [13-16]. A total of 1355 samples was tested by single donor testing and two samples were positive. However, the WNV titres were extremely low in these samples and using the in-house assay, only 1:40 and 1:41 replicates of the samples were found positive. Both samples were WNV IgM positive using a commercial assay (Panbio assay). No WNV positive donations were found in Saskatchewan after 2nd September 2003 and mini-pool testing was re-introduced in this area on 23rd September 2003.

WNV donor testing in the U.S.A. – current status

Donor testing started in the U.S.A. at the beginning of July 2003 and by September 16th approximately two million donations had been tested, of which about 490 had been interdicted from asymptomatic donors with confirmed or suspected WNV infections [17]. In the U.S.A., samples are screened under a NAT IND using a mini-pool format. Testing has effectively removed approximately 75% of infected blood donations. Mini-pool NAT testing, however, may not be able to detect low-level viraemia. A prospective study using individual donor NAT was started in high incidence areas based on the capacity for additional testing and frequency of mini-pool NAT positive units. In addition, frozen plasma was withdrawn in regions with a high incidence of WNV (as determined by mini pool NAT). Currently, the CDC is investigating two possible cases of transfusion-transmitted WNV (17).

Conclusions

Mini pool NAT testing for WNV was introduced in the U.S.A. and Canada in July this year. Testing has successfully interdicted WNV positive donations. Compared with HCV and HIV, the titres appear to be low (between 10^3 to 10^5 copies/ml). Further evaluation needs to be done on individual donor NAT testing compared with mini-pool NAT testing. Infectivity studies, particularly on the low-titre donations, would be useful. Finally, the widespread use of different NAT assays for testing blood donations in the U.S.A. and Canada makes standardisation of these methods a priority.

References

1 Petersen LR, Marfin AA, Gubler DJ : West Nile virus. JAMA 2003;290:524-528.

2 Campbell GL, Marfin AA, Lanciotti RS, Gubler DJ : West Nile virus. Lancet Infectious Diseases 2002;2:519-529.

3 Southam CM, Moore AE: Induced virus infection in man by the Egypt isolates of the West Nile virus. American Journal of Tropical Medicine and Hygiene 1954;3:19-50.

4 Biggerstaff BD, Petersen LR: Estimated risk of West Nile virus transmission through blood transfusion during an epidemic in Queens, New York City. Transfusion 2002;42:1019-1026.

5 Nash D, Mostashari F, Fine A, Miller J, O'Leary D, Murray K, Huang A, Rosenberg A, Greenberg A, Sherman M, Wong S, Layton M: 1999 West Nile Outbreak Response Working Group. The outbreak of West Nile virus infection in the New York City area in 1999. New England Journal of Medicine 2001;344:1807-1814.

6 Outbreak of West Nile-like viral encephalitis – New York 1999. MMWR Morbidity and Mortality Weekly Report 1999;48:845-859.

7 Update: investigations of West Nile virus infections in recipients of organ transplantation and blood transfusion. MMWR Morbidity and Mortality Weekly Report 2002;51:833-836.

8 Harrington T, Kuehnert HK, Lanciotti RS, Hand S, Currier M, Chamberland ME, Petersen LR, Marfin AA: West Nile virus infection transmitted by blood transfusion. Transfusion 2003;43:1018-1024.

9 Pealer LN, Marfin AA, Petersen LR, Lanciotti RS, Page PL, Stramer SL, Stobierski MG, Signs K, Newman B, Kapoor H, Goodman JL, Chamberland ME, for West Nile Virus Transmission Investigation Team: Transmission of West Nile virus through blood transfusion in the United States in 2002. The New England Journal of Medicine 2003;349:1236-1245.

10 Iwamoto M, Jernigan DB, Guasch A, Trepka MJ, Blackmore CG, Hellinger WC, Pham SM, Zaki S, Lanciotti RS, Lance-Parker SE, DiazGranados CA, Winquist AG, Perlino CA, Wiersma S, Hillyer KL, Goodman JL, Marfin AA, Chamberland ME, Petersen LR, for West Nile Virus in Transplant Recipients Investigation Team: Transmission of West Nile Virus from an organ donor to four transplant recipients. The New England Journal of Medicine 2003;348:2196-2203.

11 www.cdc.gov/ncidod/dvbid/westnile/sur&controlCaseCount03.htm.

12 Lanciotti RS, Kerst AJ, Nasci RS, Godsey MS, Mitchell CJ, Savage HM, Komar N, Panella NA, Allen BC, Volpe KE, Davis BS, Roehrig JT: Rapid detection of West Nile virus from human clinical specimens, field-collected mosquitoes and avian samples by a TaqMan reverse transcriptase-PCR assay. Journal of Clinical Microbiology 2000;38:4066-4071.

13 http://www.gov.mb.ca/health/wnv/stats.html#indicators.

14 www.health.gov.ab.ca/healthier/disease/west nile testing.html.

15 www.health.gov.sk.ca/rr wnv testresults.html.

16 www.hc-sc.gc.ca/pphb-dgspsp/wnv/mon_ehtml#sitrep.

17 Update: detection of West Nile virus in blood donations – Unites States, 2003. MMWR Morbidity and Mortality Weekly Report 2003;52:769-771.

Dr. J. Saldanha, Canadian Blood Services, 1800 Alta Vista, Ottawa, ON, K1G 4J5, Canada.
E-mail: john.saldanha@bloodservices.ca

Robertson J (ed): PDA/EMEA European Virus Safety Forum.
Dev Biol. Basel, Karger, 2004, vol 118, pp 71-77.

Variants of B19

K.E. Brown

Hematology Branch, National Heart, Lung and Blood Institute, Bethesda, MD, USA

Abstract: The human parvovirus B19 (B19), the only known human pathogenic parvovirus, was discovered in 1975. Like other parvoviruses it is a small (22-24 nm diameter) non-enveloped icosahedral virus, with a single-stranded genome. As part of its life history it replicates to high titre in the bone marrow, and in acute infection high titres of infectious virus are present in the peripheral blood. Parvovirus B19 is therefore not an uncommon 'contaminant' of blood and blood products [1]. In addition the lack of lipid membrane envelope renders parvoviruses insensitive to solvent detergents, and the small genome confers relative resistance to heat and gamma irradiation. Many medical reviews have been written on the molecular biology, clinical features, approaches to diagnosis and management, and these should be consulted for more information [2]. This review will give an overview of the biology of parvovirus B19 infection, and the more recently described variants of parvovirus B19.

TAXONOMY OF THE *PARVOVIRIDAE*

The *Parvoviridae* are divided into two sub-families, *Parvovirinae* and *Densovirinae*, on the basis of their ability to infect vertebrate or invertebrate cells respectively. The *Parvovirinae* are further sub-divided into three genera on the basis of their transcription maps and the ability to replicate either autonomously (genus *Parvovirus*), with helper virus (genus *Dependovirus*), or in erythroid progenitor cells (genus *Erythrovirus*) [3]. Bovine parvovirus, canine parvovirus and porcine parvovirus are members of the Parvovirus genus, and are readily cultured in vitro in a variety of different cell lines. Parvovirus B19 only efficiently replicates in erythroid progenitor cells, the cells that produce the circulating red cells, and is the type member of the Erythrovirus genus. Other members of the Erythrovirus have since been described, including the simian parvovirus (SPV) that infects cynomolgus monkeys, the rhesus parvovirus and the pig-tailed macaque parvovirus.

Parvovirus B19 obtains its name from its original observation in a sample designated 'B19' in sera tested for hepatitis B by counter-immune electrophoresis (CIE). When the precipitin line was examined by electron microscopy, parvovirus-like particles were observed, and the virus was originally given the name human

SPLV (serum parvovirus-like virus). Subsequently the viral genome was characterized and the parvovirus nature was confirmed. However, the accronym HPV was already associated with human papillomavirus, and to avoid confusion the new virus was named parvovirus B19.

Viral Genome

Parvovirus B19 is a typical member of the Parvoviridae. The entire virus has been cloned and sequenced. The virus has a lined single-stranded genome of ~ 5,596 nucleotides, and the viral capsids encapsidate a single strand of either negative or positive polarity. The 5' and 3' ends have identical terminal repeat sequences 383 nucleotides in length, with hairpin structures of 176 nucleotides. These hairpin structures are considerably longer than in other parvoviruses, and until recently [4] the full-length viral genome had not been successfully cloned into a bacterial plasmid, and no infectious clone had been produced.

The viral genome is highly conserved, with generally only 1-2% sequence variation between isolates. However, recently variants with ~ 10% sequence variation at the nucleotide level have been described (see below).

Parvovirus B19 only replicates in erythroid progenitor cells, and therefore has been difficult to grow in culture. The virus will replicate in erythroid progenitors obtained from human bone marrow, peripheral blood or foetal liver. However all these culture systems are primary cultures only, with the erythroid cells only surviving for < 10 days. The virus can also be propagated in a few rare cell lines, including the erythropoietin-dependent cell lines UT7/Epo, and its sub-clone, UT7/Epo-S1 and KU812Ep6 cells.

B19 Proteins

The B19 genome has a single promoter (p6) and two large open reading frames (ORFs). The large non-structural protein, NS1, is encoded from a non-spliced transcript from the ORF on the left-hand side of the genome. The two capsid proteins are encoded in the right hand side of the genome by two alternatively spliced transcripts. The virion comprises 60 copies of the capsid proteins, of which ~ 95% are the major capsid protein, VP2 with a molecular weight of 58k Da. The remaining 5% of the virion is VP1, which has the same structure as VP2, but with an additional 226 amino acids at the amino terminus.

The capsid proteins can be expressed in insect cells using the baculovirus expression system. Expression of VP2 alone leads to the self-assembly of virus-like particles, which appear morphologically and antigenically identical to native virions. VP2 only capsids bind to red cells, through interaction with the parvovirus B19 receptor, globoside, also known as blood group P antigen. In contrast, VP1 is not required for interaction with the receptor, and will not self-assemble to form viral-like capsids. However the 226 amino-acid sequence that is not shared with VP2 (the VP1-unique region) is the main site of neutralizing epitopes, and is necessary to produce a neutralizing antibody response.

B19 Epidemiology

Parvovirus B19 is a common infection, with a worldwide distribution. Seroconversion occurs during childhood and early adult life, and > 50% of adults have antibody to parvovirus B19. It has been estimated that the annual seroconversion rate of women in the U.S.A. is ~ 1.5%. Outbreaks of parvovirus B19 infection are not uncommon, and in temperate countries B19 infection is more common during the spring months.

The main route of transmission is via the respiratory system. However, the intense viraemia that occurs in early infection, with titres of ~ 10^{12} genome copies/ml or greater, means that the virus can be transmitted by blood and blood products. This is particularly common in pooled products (clotting factors and plasma), but has also been documented more rarely following administration of single components (red cells and platelets) [1]. The prevalence of B19 in blood donors depends on the method chosen (sensitivity) to detect B19 and the time of year. Studies of the prevalence of B19 in blood donors indicate that ~ 1:20,000 donors have B19 titres > 10^9 genome copies/ml, with ~ 1:1,000 donors having B19 detected by more sensitive methods such as PCR.

Time Course of Parvovirus B19 Infection

The kinetics of B19 infection was first determined by the production of infection in healthy volunteers. Following intranasal inoculation, significant levels of viraemia (>10^6 genome copies/ml) were detected at days 5-6, with a peak of ~ 10^{12} at days 8-9, and then an equally rapid drop to < 10^6 genome copies/ml. B19-specific IgM antibody appeared about 10-12 days after experimental inoculation and B19-specific IgG antibody appeared at about two weeks, and correlated with the onset of a rash like-illness and arthropathy, the clinical symptoms of fifth disease. The time course is similar in natural infections, with IgM detectable for approximately three months following infection and IgG detectable life-long thereafter.

Clinical Manifestations of B19 Infection

Although many infections are asymptomatic, the clinical signs and symptoms of parvovirus B19 infection vary widely with the haematological and immunological status of the host, with haematological disease due to direct viral cytotoxicity in erythroid progenitors, and the rash and arthropathy due to immune-complex mediated disease (Table 1). Most infection is probably asymptomatic, but the commonest presentation is as fifth disease (erythema infectiosum, or slapped cheek disease), a common innocuous disease of childhood. Individuals initially develop a florid rash over the cheeks, before the appearance of a lacy rash over the rest of the body, especially the arms and legs. The illness is generally benign, but in some individuals, and especially in women, it may be associated with joint pains and even arthritis. The symptoms may mimic acute rheumatoid arthritis, with the arthritic symptoms lasting several weeks before resolution. The association with chronic arthropathy is controversial, as B19 DNA remains detectable in synovial tissues for years following B19 infection, and can be detected in both individuals with and without joint symptoms.

Table 1: Disease manifestations and persistence of parvovirus B19 infection in different host populations.

Disease	Acute/Chronic	Host
Fifth disease	Acute	Normal children
Polyarthropathy syndrome	Acute/chronic	Normal adults
Transient aplastic crisis	Acute	Patients with increased erythropoiesis
Hydrops foetalis/ congenital anaemia	Acute/Chronic	Foetus (<20 weeks)
Persistent anaemia	Chronic	Immunodeficient /immunocompromised patients

In patients with increased red cell turnover, (e.g. patients with sickle cell disease or haemolytic anaemia), B19 infection manifests as transient aplastic crisis, a precipitous worsening of anaemia due to the transient cessation of red cell production. Although this can be life-threatening without medical intervention, as the name suggests, it is only transient, and red cell production restarts as the virus is neutralized by the immune response. However, in individuals who are unable to mount an effective immune response (patients with immunodifficiencies, including some AIDS patients), the virus is not cleared and patients develop chronic anaemia due to the persistent infection. Such patients can be successfully treated with commercial intravenous immunoglobulin which contains high levels of anti-B19 neutralizing antibodies.

B19 infection during pregnancy, especially in the mid-trimester at the time of rapid increase in the red cell mass, may lead to spontaneous abortion or the development of foetal hydrops. In addition, rare cases of congenital anaemia have been associated with B19 in pregnancy.

Many other diseases have been reported to be associated with parvovirus B19 infection, mainly based on the detection of B19 DNA in tissues by PCR in case studies or limited series of samples without controls (Table 2).

Table 2 : Possible disease associations with parvovirus B19.

Paroxysmal cold haemoglobulinuria
Haemophagocytic syndrome
Idiopathic thrombocytopenia
Pancytopenia
Necrotizing vasculitis
Kawasaki disease
Sclerosing glomerulonephritis
Hepatitis
Myocarditis
Encephalitis/chronic fatigue syndrome
Rheumatoid arthritis
Systemic lupus

Primate Parvoviruses

Parvovirus B19 is not the only member of the Erythrovirus genus: a fatal outbreak of anaemia in cynomolgus monkeys at Wake Forest USA led to the identification of the simian parvovirus, SPV [5]. Five animals were group-housed, and four of the animals became anaemic. All the animals had evidence of infection by a simian retrovirus, but the four anaemic monkeys also had evidence of erythroviral-like sequences in their serum, and parvoviral particles were seen in the bone marrow of one of the monkeys that died. The virus was characterized, and the entire coding region cloned and sequenced [6].

SPV has little homology with the autonomous parvoviruses or the dependoviruses, but 50% overall homology with parvovirus B19 DNA. At the amino acid level there was 70% homology with B19 capsid proteins and 50% homology with B19 non-structural protein. Like B19, SPV selectively infects erythroid progenitor cells, and inoculation into naive cynomolgus macaques produces disease very similar to B19 infection in humans [7], including the production of foetal hydrops in pregnant macaques (O'Sullivan et al, manuscript in preparation). Similar parvoviruses have been identified in other macaques [8], any of which could be used as animal models of B19 infection.

Identification of B19 Variants

Until recently, all the B19 isolates had very similar genomic sequence, with less than 5% sequence variation. However, in 1998 a French group reported the identification of a B19 sequence which was amplified by B19 PCR primers, but did not hybridize with a B19 specific probe [9]. The viral genome was cloned and sequenced, and confirmed that the isolate differed from other B19 sequences by 11-14%, and that standard B19 PCR assays would not detect the variants [10]. However, most of the nucleotide differences were at the third nucleotide position or silent mutations, and the viral proteins have 96-97% homology with other B19 isolates [11]. The capsid proteins were expressed in insect cells and, like B19, self-assemble to form viral capsids with similar antigenicity [11] and binding to globoside (Heegaard and Brown, unpublished observations) as B19 capsids.

Identification of a Second Group of Variant B19 Sequences

As part of our studies of parvovirus B19, and as a source of samples for looking for infectious virus, identifying novel viral strains, and developing new diagnostic assays, we offer diagnostic testing for B19 DNA detection for NIH and outside institutions. With the original publication of V9 we looked for similar sequences in these samples using the PCR described in their manuscript [10]. When Dr. Nguyen joined the laboratory, we were able to obtain both a clone of the V9 sequence, and the full V9 sequence which had not been published at the time. Although our B19 DNA screening assay, dotblot hybridization, would detect V9 sequences, our standard PCR would miss low titre samples. To detect V9 and similar sequences we used a new PCR assay whose primers were designed from a conserved region of the

B19/V9 sequence, and tested samples submitted to the laboratory for B19 testing. We identified one variant sequence (~ 90% similarity to both B19 and V9) in a sample coded A6, from an anaemic HIV positive patient from Italy [12]. As with V9, most of the variation was at the third nucleotide position, and the putative capsid proteins sequence had 96-98% similarity to other B19 sequences. The sequence similarity at the amino acid level between A6, V9 and B19 suggests that these viruses should have the same serotype, but without infectious virus this hypothesis has not been tested.

In similar studies a Finnish group studied the prevalence of B19 in skin tissues and found a mis-match between the results of testing for B19 DNA with different primer pairs. Subsequent sequencing of the products led to the identification of a B19 genotype which they named K71 and which differed from the published B19 and V9 sequences [13], and was almost identical to the A6 genotype. All the patients with K71 sequence detected in their skin tissue had B19 detectable antibodies, again suggesting that there is no serological difference between K71 and B19. Similar sequences have also been detected in liver samples from the U.S.A. [14], and serum samples from France [15].

Prevalence of B19 Variants

The sequence variation between B19 and the variants are scattered throughout the genome, and unless primers are specifically designed to conserved regions between the variants, standard B19 primers will miss the variants, almost certainly accounting for why these sequences have only recently been detected. Using such primers, a number of preliminary studies to determine the prevalence of the variants have been performed. In collaboration with Erik Heegaard we tested plasma pools from Danish blood donors. Each pool contained plasma from 2,000 individuals. Although B19 sequences were detected in 40/62 pools, no variant sequences were identified [12]. Similarly, in a French study, B19 sequences were detected in 204/270 serum samples obtained from U.S.A. individuals with B19-like symptoms, but no variants were found [15]. In contrast, B19 sequences were detected in 88/633 sera from French individuals with B19-like symptoms, nine of which were variant sequences. In liver tissues we identified that 7/37 samples with detectable B19 sequences were variant sequences. In the skin studies, 5/14 of the detectable sequences were variant sequences.

Conclusions

Since its discovery in 1974 much has been learnt about the biology of parvovirus B19. Although originally thought to have a single genotype, the identification of isolates with ~ 10% sequence variation suggests the prevalence of three discrete genotypes of human parvovirus B19. These genotypes can be detected in human sera and tissues if the appropriate primers are used. However the true prevalence of these new genotypes is still unknown, and may differ in different countries. Similarly, the clinical significance of these viruses is unknown. Current PCR-based assays are unlikely to detect variants unless the primers are based on conserved sequences between the three different genotypes.

References

1 Azzi A, Morfini M, Mannucci PM: The transfusion-associated transmission of parvovirus B19. Transfus Med Rev 1999;13:194-204.

2 Heegaard ED, Brown KE: Human parvovirus B19. Clin Microbiol Rev 2002;15:485-505.

3 Berns KI, Bergoin M, Bloom M, Lederman M, Muzyczka N, Siegl G et al: Family Parvoviridae, in Regenmortel MHV, Fauquet CM, Bishop DHL, Carstens EB, Estes MK, Lemon SM et al (ed): Virus Taxonomy: Classification and Nomenclature of Viruses: Seventh Report of the International Committee on Taxonomy of Viruses. New York, Academic Press, 2000, pp 169-178.

4 Zhi N, Zadori Z, Brown KE, Tijssen P: Construction and sequencing of an infectious clone of the human parvovirus, B19. Virology 2004;318:142-152.

5 O'Sullivan MG, Anderson DC, Fikes JD, Bain FT, Carlson CS, Green SW et al: Identification of a novel simian parvovirus in cynomolgus monkeys with severe anemia: a paradigm of human B19 parvovirus infection. J Clin Invest 1994;93:1571-1576.

6 Brown KE, Green SW, O'Sullivan MG, Young NS: Cloning and sequencing of the simian parvovirus genome. Virology 1995;210:314-322.

7 O'Sullivan MG, Anderson DK, Goodrich JA, Tulli H, Green SW, Young NS et al: Experimental infection of cynomolgus monkeys with simian parvovirus. J Virol 1997;71:4517-4521.

8 Green SW, Malkovska I, O'Sullivan MG, Brown KE: Rhesus and pig-tailed macaque parvoviruses: identification of two new members of the Erythrovirus genus in monkeys. Virology 2000;269:105-112.

9 Nguyen QT, Sifer C, Schneider V, Bernaudin F, Auguste V, Garbarg-Chenon A: Detection of an erythrovirus sequence distinct from B19 in a child with acute anaemia [letter]. Lancet 1998;352:1524.

10 Nguyen QT, Sifer C, Schneider V, Allaume X, Servant A, Bernaudin F et al: Novel human erythrovirus associated with transient aplastic anemia. J Clin Microbiol 1999;37:2483-2487.

11 Heegaard ED, Qvortrup K, Christensen J: Baculovirus expression of erythrovirus V9 capsids and screening by ELISA: serologic cross-reactivity with erythrovirus B19. J Med Virol 2002;66:246-252.

12 Nguyen QT, Wong S, Heegaard ED, Brown KE: Identification and characterization of a second novel human erythrovirus variant, A6. Virology 2002;301:374-380.

13 Hokynar K, Soderlund-Venermo M, Pesonen M, Ranki A, Kiviluoto O, Partio EK et al: A new parvovirus genotype persistent in human skin. Virology 2002;302:224-228.

14 Wong S, Young NS, Brown KE: Prevalence of parvovirus B19 in liver tissue: no association with fulminant hepatitis or hepatitis-associated aplastic anemia. J Infect Dis 2003;187:1581-1586.

15 Servant A, Laperche S, Lallemand F, Marinho V, De Saint MG, Meritet JF et al: Genetic diversity within human erythroviruses: identification of three genotypes. J Virol 2002;76:9124-9134.

Dr. K.E. Brown, National Institutes of Health, Building 10 / Room 7C218, 9000 Rockville Pike, Bethesda, MD 20892, USA
E-mail: brownk@nhlbi.nih.gov

Robertson J (ed): PDA/EMEA European Virus Safety Forum.
Dev Biol. Basel, Karger, 2004, vol 118, p 79.

SESSION III (1)

Virus Validation, Virus Assay and Standardisation

Chairs: *H. Willkommen* (Clearant)
G. Rössling (Schering)

Robertson J (ed): PDA/EMEA European Virus Safety Forum.
Dev Biol. Basel, Karger, 2004, vol 118, pp 81-88.

Quantitative NAT for Pathogen Inactivation Verification

K. McKenney, L. Gillmeister, K. Marlowe, D. Armistead, W. Burgess, W. Drohan

Clearant Inc., Gaithersburg, MD, USA

Key words: Pathogen inactivation, quantification, Q-NAT, LTQ-NAT

Abstract: A novel Quantitative Nucleic Acid Test (Q-NAT) technology has been developed to demonstrate, quantify and verify pathogen inactivation by methods that break pathogen nucleic acids, specifically, gamma irradiation. The Q-NAT technology provides significant advantages in cost, efficiency and broad applicability compared with traditional methods for pathogen inactivation detection and quantification such as cell culture.

INTRODUCTION

Detection and quantification of pathogen contamination in biological products is essential to provide safe products and successful processes. Not all pathogens are readily detectable or quantifiable by bioassay, nor are all pathogens readily inactivated. Clearant Inc. has developed the "Clearant Process", a comprehensive method of inactivating pathogens in biological products as well as hard and soft tissue using gamma radiation while preserving product activity, structure and function [1-3].

Current 'state of the art' Quantitative NAT (Q-NAT) technology uses a small target amplicon (~100 bp) to detect and quantify pathogens in blood and blood products [4-9]. We have developed a new application of Q-NAT technology with large target amplicons (LTAs, ~0.5-8.0 Kbp) that, in conjunction with small target amplicons, provides a direct, quantifiable measure of both the pathogen's presence and the pathogen's inactivation.

LTQ-NAT technology for pathogen inactivation could potentially significantly reduce costs associated with bioassays, where available, and prove especially useful where bioassays are not available (e.g. Hepatitis B, (HBV), and Hepatitis C (HCV) or where bioassays are difficult to perform (eg. human B19 parvovirus). Plasma-derived product 'look back' costs could be dramatically reduced if the LTQ-NAT technology can clearly demonstrate pathogen nucleic acid is present but not in a form that can be replicated. The LTQ-NAT technology could provide significant benefit by considerably reducing costs of product safety monitoring, process development and validation.

The development of LTQ-NAT is presented in the context of gamma irradiation-induced fragmentation of cloned viral pathogen target molecules as well as bona fide viral pathogens. LTQ-NAT in conjunction with end point PCR results are presented. The experimental framework for comparing NAT determination of pathogen inactivation with bioassay determination of pathogen inactivation is described.

Materials and Methods

Viral pathogens and molecular clones

Molecular clones of near full length B19 virus were constructed from positive plasma. Gamma irradiation-induced fragmentation experiments were conducted with cloned non-infectious human B19 parvovirus that had been irradiated to 50 kGy at –80 C in (type of gamma irradiator). In addition, B19 and HBV positive plasma were obtained commercially, irradiated to 50 kGy and used in the experiments described.

LTQ-NAT: reactions, primers and probes

Q-NAT was performed on an ABI 7000 instrument (Applied Biosystems, Foster City, CA). Primers and probes are designed with 'Primer Express' software using pathogen sequences obtained from GenBank, literature references and cloned sequences determined by Clearant. The LTQ-NAT data have been generated using Taqman probes with Fam and Tamra dyes as well as MGB probes from ABI.

Gamma irradiation

The irradiation dose for B19 virus and clones was 50 kGy with the Clearant Process. The Clearant Process was developed for pathogen inactivation in plasma and tissue and uses a number of important process conditions including dose rate, temperature and radio protectants.

Irradiation was performed at the NIST "gamma cell", a ^{60}Co irradiator.

Results and Discussion

Our initial work with LTQ-NAT used a non-infectious molecular clone of the human B19 parvovirus as a model template to develop a 'Taqman' assay for quantitative PCR analysis. Figure 1 shows an experiment with a small target amplicon (119 bp) using the B19 clone over a range of concentrations from 10^{11} to 10^{1}genome equivalents/reaction.

The data in Figure 1 are displayed as a linear fluorescent signal on the Y axis versus the PCR cycle number on the X axis. The results demonstrate a reasonable signal over the range of 10^{11} to 10^{4} genome equivalents per reaction. The lower limit of sensitivity appears to be between 10^{4} and 10^{3} molecules of template per reaction, suggesting the reaction is not optimal.

The data in Figure 1 were used to generate a plot of template concentration versus cycle number of fluorescence, shown in Figure 2. The plot shows a fairly typical quantitative PCR 'Taqman' assay result for a small target amplicon using a cloned DNA template. The challenge in the development of the LTA-Q-PCR assay is implementing large target amplicons (LTA) to detect and quantify loss of the pathogen genomes.

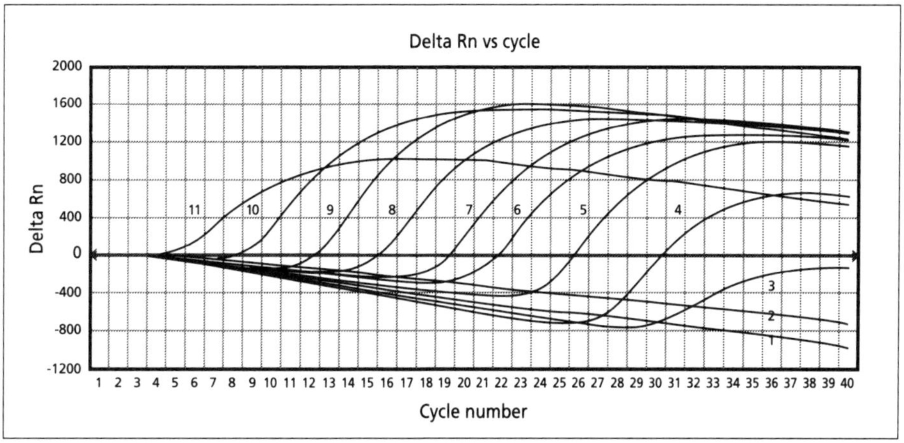

Fig. 1: Q-PCR Taqman assay with a 119bp amplicon on a cloned human B19 virus template.

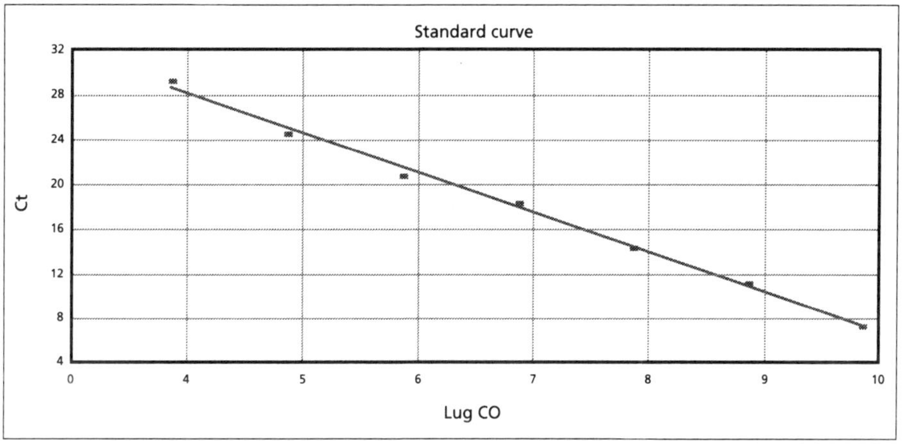

Fig. 2: Plot of the cycle number versus log concentration of DNA template from Figure 1.

LTQ-PCR with large target amplicons

Large PCR amplicons have been used in a number of applications over the past ten years including cloning large fragments, site directed mutagenesis and mapping, among others. The challenge in developing the large target amplicon Q-NAT technology is not making the large amplicons, but rather, making large amplicons under conditions that work for quantitative PCR instruments. Figure 3 shows results using two different size amplicons, one of 80 bp (left most set of four profiles) and one of 697 bp (right most set of two profiles) to detect the HBV pathogen under 'standard' Taqman assay conditions. The figure demonstrates that the small amplicon works reasonably well, while the larger amplicon works poorly. This loss of large amplicon efficiency has been seen by a number of Q-PCR experimentalists.

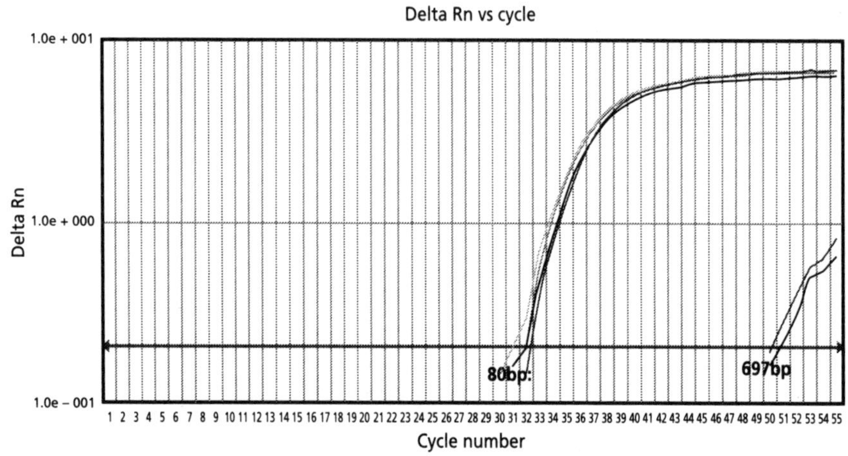

Fig. 3: Effect of amplicon size on the Q-PCR Taqman assay for an 80bp amplicon (left @32 cycles) and a 697bp amplicon (right @ 51 cycles).

To develop a robust LTQ-PCR assay, we first used an endpoint PCR assay to confirm the synthesis of products. The endpoint assays do not use a probe and provide a 'snapshot' of the overall reaction at the end of a fixed number of cycles of amplification. This permits a number of different reaction conditions, primer combinations and amplicon sizes to be explored. Figure 4 shows the results of an endpoint PCR assay for a B19 positive plasma sample with small (STA, 112 bp) and large (LTA, 2.4 Kbp) B19 amplicons after forty cycles of amplification. In addition, a sample of the plasma has been irradiated at 50 kGy to inactivate pathogens under conditions that preserve greater than 85% of the structure and function of the albumin in the sample.

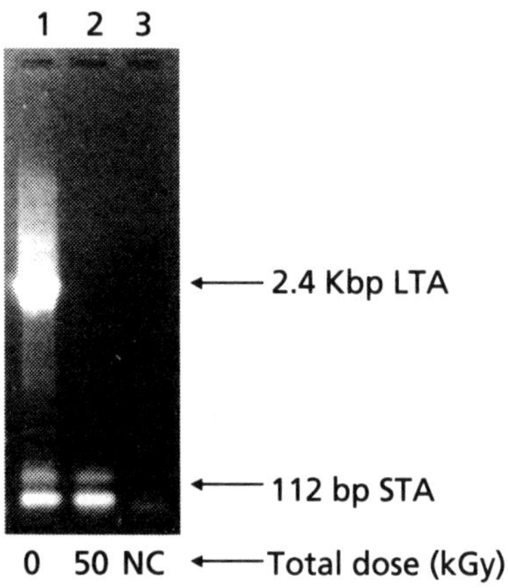

Fig. 4: Multiplex PCR endpoint assay for human B19 virus at 0 & 50 kGy of gamma irradiation with a 2.4 Kbp large target amplicon and 112 bp small target amplicon.

The data presented in Figure 4 demonstrates a qualitative loss of the large target amplicon (2.4 Kbp) of the B19 virus after 50 kGy of irradiation with little or no loss of the small target amplicon (STA, 112 bp). The large amplicon in this experiment represents approximately 45% of the B19 virus genome.

The endpoint PCR assays provide a direct, albeit qualitative, analysis of the destruction of the pathogen nucleic acid. To develop a quantitative assay with large amplicons, we have modified the Q-PCR Taqman assay conditions. The new LTQ-PCR reaction conditions have modified polymerases (to include a 3' proofreading activity), salts, annealing and extension times and temperatures, as well as modified nucleotide and primer/probe concentrations relative to standard Taqman assay conditions.

One application of the LTQ-PCR is shown in Figures 5-7. The template used in the assay shown is a cloned human B19 virus that has been irradiated to 50 kGy under conditions that preserve proteins and tissues. Figure 5 shows the small target amplicon analysis of B19 clone exposed to 50kGy and is similar to the data shown in Figure 1 for the B19 clone that has not been irradiated. Figure 6 shows the large target amplicon (2.4 Kbp) analysis of the non-irradiated (0 kGy) cloned DNA, and Figure 7 shows the large target amplicon analysis of the 50 kGy irradiated cloned DNA. Figure 7 shows a significant cycle delay in the appearance of the large target amplicon with 50 kGy irradiated DNA.

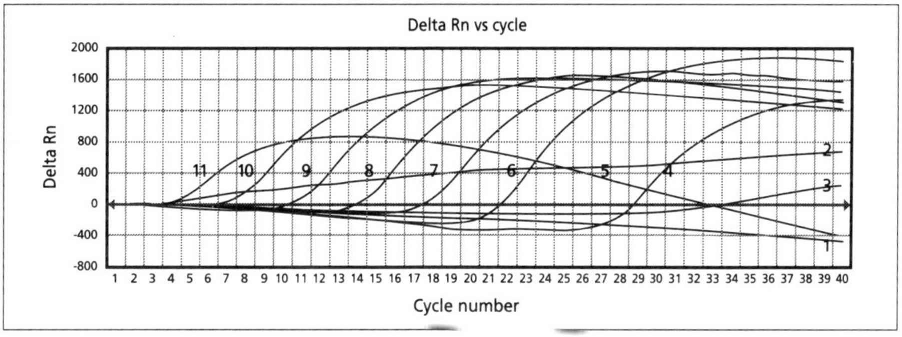

Fig. 5: Small target amplicon (119 bp) Q-PCR analysis of 10-fold dilutions of B19 cloned DNA exposed to 50 kGy. The number on each line refers to the number of genome equivalents in each reaction

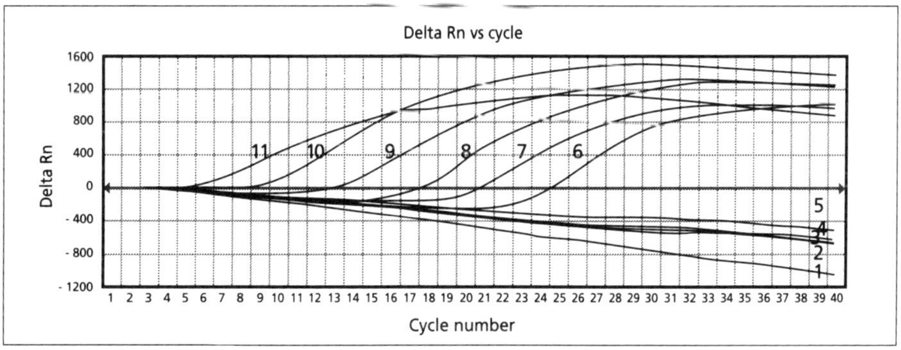

Fig. 6: Large target amplicon (2.4 Kbp) Q-PCR analysis of 10-fold dilutions of B19 cloned DNA. The number on each line refers to the number of genome equivalents in each reaction.

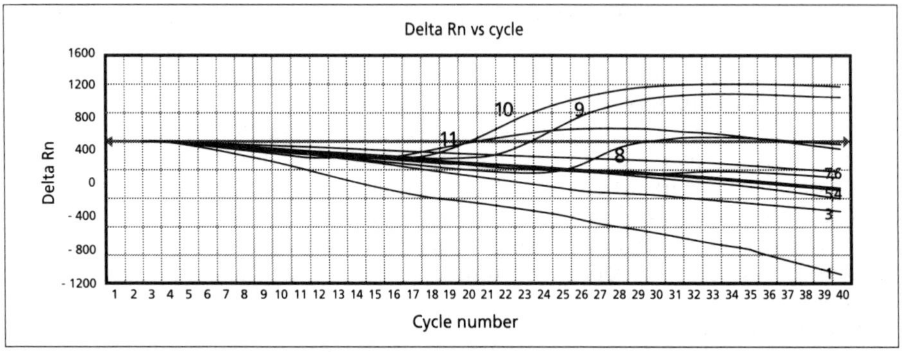

Fig. 7: Large target amplicon (2.4 Kbp) Q-PCR analysis of 10 fold dilutions of B19 cloned DNA irradiated to 50 kGy. The number on each line refers to the number of genome equivalents in each reaction.

Quantification of pathogen inactivation using LTQ-PCR

We would ideally use an LTA that corresponds to the full genome size of the viral pathogen of interest. However, it is possible that for technical reasons, we may not be able to make a full-length viral amplicon for all viral pathogens. As an alternative to the direct quantification of a full-length amplicon, we can estimate the loss of the full length amplicon from a plot of a number of smaller amplicons. As an example, we have plotted the data from Figures 5-7 in Figure 8. The data in Figure 8 display the C_T or threshold cycle number, on the Y axis, versus the percentage of

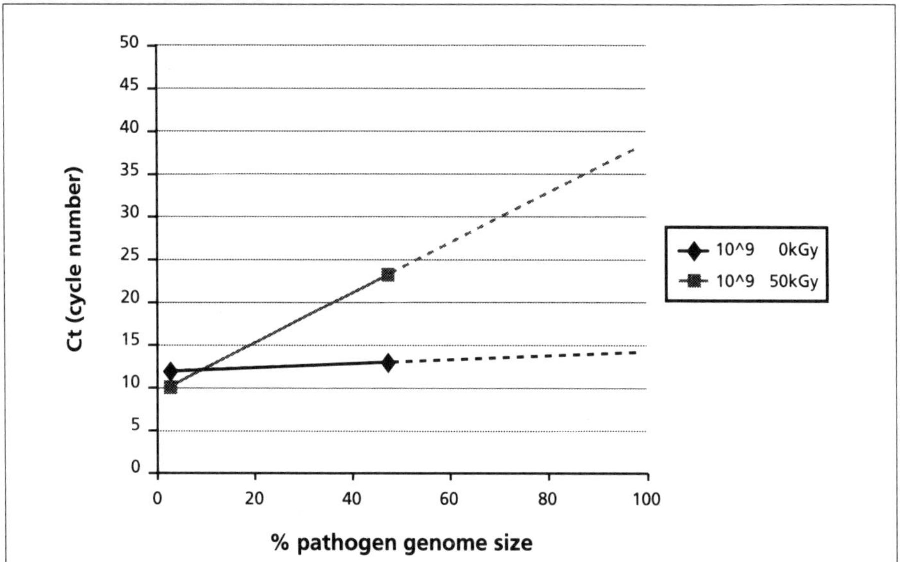

Fig. 8: Plot of C_T versus amplicon size as a % of the pathogen genome.

the pathogen's genome size represented in the target amplicon. The small amplicon in this example is 2% of the B19 viral genome, while the large 2.4 Kbp amplicon is 45% of the B19 viral genome. If we examine the data at one concentration (eg 10^9), there is little difference between the C_T for the small and large amplicons (12 vs. 13 cycles) and for the estimated full length pathogen amplicon (14 cycles) at 0 kGy. After irradiation to 50 kGy, there is a significant difference in the C_T between the small and large amplicons (10 vs. 24) and an even larger difference for the estimated full-length amplicon (38 cycles). The delay between the estimated full-length (38 cycles –14 cycles) is approximately 24 cycles. The reaction efficiency can be determined from ten-fold dilution curves for the large amplicon to be approximately 4.0 cycles to make a ten-fold increase in product or stated another way, the 24 cycle delay divided by the 4.0 cycles/10 fold difference in concentrations suggests the loss of 6 logs of full length B19 cloned DNA as a direct result of the exposure to 50 kGy of gamma irradiation.

Conclusions

The estimates of the pathogen inactivation provided above are meant to illustrate a new experimental approach to the quantification of pathogen inactivation by methods that introduce damage in a pathogen's DNA or RNA. Pathogen inactivation resulting from gamma irradiation provides an ideal substrate for the development of this approach. Other pathogen inactivation methods may be candidates for this approach if the 'inactivation' mechanism introduces strand breaks directly, or results in changes to the pathogen's genetic material so that it can be accessible to RNA or DNA modifying enzymes before the LTQ-NAT assay. Human Hepatitis B virus LTQ-NAT assays have been developed and provide results similar to the B19 cloned DNA results provided above.

References

1 Grieb T, Forng RY, Brown R, Owolabi T, Maddox E, McBain A, Drohan WN, Mann DM, Burgess WH: Effective use of gamma irradiation for pathogen inactivation of monoclonal antibody preparations. Biologicals 2002;30(3):207-216.

2 Miekka SI, Forng RY, Rohwer RG, MacAuley C, Stafford RE, Flack SL, MacPhee M, Kent RS, Drohan WN: Inactivation of viral and prion pathogens by gamma-irradiation under conditions that maintain the integrity of human albumin. Vox Sang 2003;84(1):36-44.

3 Miekka SI, Busby TF, Reid B, Pollock R, Ralston A, Drohan WN: New methods for inactivation of lipid-enveloped and non-enveloped viruses. Haemophilia 1998;4(4):402-408.

4 Pas SD, Fries E, De Man RA, Osterhaus AD, Niesters HG: Development of a quantitative real-time detection assay for hepatitis B virus DNA and comparison with two commercial assays. J Clin Microbiol 2000;38(8):2897-2901.

5 Loeb KR, Jerome KR, Goddard J, Huang M, Cent A, Corey L: High-throughput quantitative analysis of hepatitis B virus DNA in serum using the TaqMan fluorogenic detection system. Hepatology 2000;32(3):626-629.

6 Chen RW, Piiparinen H, Seppanen M, Koskela P, Sarna S, Lappalainen M: Real-time PCR for detection and quantitation of hepatitis B virus DNA. J Med Virol 2001;65(2):250-256.

7 Abe A, Inoue K, Tanaka T, Kato J, Kajiyama N, Kawaguchi R, Tanaka S, Yoshiba M, Kohara M: Quantitation of hepatitis B virus genomic DNA by real-time detection PCR. J Clin Microbiol 1999;37(9):2899-2903.

8 Desire N, Dehee A, Schneider V, Jacomet C, Goujon C, Girard PM, Rozenbaum W, Nicolas JC: Quantification of human immunodeficiency virus type 1 proviral load by a TaqMan real-time PCR assay. J Clin Microbiol 2001;39(4):1303-1310.

9 Albadalejo J, Alonso R, Antinozzi R, Bogard M, Bourgault AM, Colucci G, Fenner T, Petersen H, Sala E, Vincelette J, Young C: Multicenter evaluation of the COBAS AMPLICOR HCV assay, an integrated PCR system for rapid detection of hepatitis C virus RNA in the diagnostic laboratory. J Clin Microbiol 1998;36:862-865.

Dr. K. McKenney, Clearant Inc., 401 Professional Drive, Gaithersburg, MD 20879, USA.
E-mail: kmckenney@clearant.com

Robertson J (ed): PDA/EMEA European Virus Safety Forum.
Dev Biol. Basel, Karger, 2004, vol 118, pp 89-98.

Use of Bacteriophages as Surrogates for Mammalian Viruses

M. McAlister, H. Aranha, R. Larson

Scientific and Laboratory Services (SLS), Pall Corporation, Port Washington, NY, U.S.A.

Key words: Virus clearance, bacteriophage, filtration, size exclusion, validation

Abstract: The threat of viral contamination is common to all processes using biological products of animal or human origin. Therefore, demonstration of virus clearance (i.e. validation of virus removal and/or inactivation steps) is of utmost importance to the biopharmaceutical industries. Ultimately, virus clearance studies should show that any virus removal/inactivation stage incorporated into the manufacturing process not only removes or inactivates known viruses that may be conceivably present (e.g. from cell banks and source materials), but also other viruses that may be introduced adventitiously (e.g. by addition of supplements downstream of the manufacturing process).
In this paper, we outline the shared properties of mammalian viruses and similar sized bacteriophages, and factors that may influence the virus clearance process. We also present test data from filtration studies, showing similar titre reductions for both types of virus. We propose that well-characterised bacteriophage, such as PP7 and PR772 can be used as models for mammalian viruses if the virus removal mechanism is based on size exclusion.

INTRODUCTION

As the search for new, innovative products for future therapeutic use continues, use of products from animal or human origin seems certain to increase concurrently. These products may be biotechnology-derived (e.g. monoclonal antibodies, enzymes and recombinant proteins), plasma-derived (e.g. immunoglobulins, albumins and hormones), or derived from micro-organisms (e.g. vaccines, recombinant proteins and toxoids).

However, all products manufactured from material of human or animal origin may potentially be contaminated by animal viruses. Consequently, all manufacturing processes incorporating biological products must include at least one stage demonstrating significant virus clearance, as stipulated by the regulatory authorities [1,2].

The objective of virus clearance studies is two-fold. Firstly, the virus clearance stage must be shown to remove or inactivate viruses that may be present, either as a consequence of the product origins (e.g. presence of endogenous retroviruses in the cell bank, animal or human tissues), or resulting adventitiously during the production process (e.g. operator handling, addition of supplements) [1-4]. Secondly, the virus clearance step must demonstrate the robustness of the removal or inactivation process, using non-specific model mammalian viruses [1,2].

Successful virus validation studies require careful consideration of many factors, including selection of the model virus, the level of clearance required, and the properties of the product. In this paper, we outline these factors, and compare virus clearance data from filtration studies for both mammalian viruses and bacterophages. Based on the results from numerous studies, we propose that bacteriophages should be considered as surrogates for mammalian virus when the mechanism to effect virus clearance is based on size exclusion.

Virus Clearance: Inactivation or Removal?

To be assured of the highest level of virus clearance, there are many factors to consider before deciding whether to use a removal- or inactivation- based mechanism. However, the parameters associated with an existing production process will contribute largely to this decision, and any virus clearance step must be examined in the context of a particular production process for a particular product [2]. However, an increasing number of manufacturers are now considering a generic approach to virus clearance, provided that it can be adequately demonstrated that viral clearance is not influenced by operating parameters that vary slightly between processes [5].

Virus inactivation may occur by physical or chemical means. Commonly used physical inactivation methods include the application of heat or radiation, both of which render the virus non-infectious and non-replicating [3]. However, these treatments may adversely affect the properties of the product in an irreversible manner (e.g. protein denaturation), thereby limiting their usage.

Chemical inactivation of viruses most often involves detergent and solvent treatment. However, only enveloped viruses are inactivated by this treatment, thereby limiting the usefulness of this method. Other chemicals that have been considered include β-propiolactone, and in the early days of biotechnology, formalin was used. However, β-propiolactone reacts with proteins, thereby altering the final product, and the toxicity issues associated with formalin make this an impractical option.

The main disadvantage of virus inactivation technologies for virus clearance is the limited number of viruses affected. Viruses are classified as either enveloped or non-enveloped. This refers to the presence or absence of a lipid membrane surrounding the protein shell found in many viruses. Inactivation methods such as solvent and detergent treatment exert their inactivation effects by damaging this lipid membrane. Therefore, a second virus clearance mechanism would be required in the manufacturing process to address the potential presence of non-enveloped virus.

This problem is overcome by using a removal-based method. The main removal methods used for virus clearance on an industrial scale are filtration (direct flow filtration (DFF), tangential flow filtration (TFF)), chromatography (ion-exchange, affinity, etc) and to a lesser extent, partitioning by fractionation (partitioning is the least effective removal mechanism, but may contribute to the overall clearance reported in the manufacturing process) [1]. As well as targeting both enveloped and non-enveloped virus, the product is usually not changed by application of a removal method such as filtration [4]. In addition, some filter manufacturers have developed non-destructive, integrity tests that measure the forward flow of gas through a wetted filter under air or nitrogen pressure (diffusion through wetted pores and bulk flow through any unwetted pores). This is correlated to virus retention, thereby providing the user with a greater level of filter performance assurance [6-8]. In addition, for sterilizing filtration, the regulatory authorities state that post-use integrity tests must be completed before product release, and this can also be considered applicable to virus filtration [4].

Several virus clearance stages can be incorporated into a manufacturing process. If each stage uses a different virus clearance mechanism, then the overall clearance of the process is defined as the sum of both steps combined. This rule does not apply if two stages of the same mechanism are used (e.g. two low pH inactivation steps in series) [3]. It should be noted that a dedicated virus clearance step must demonstrate greater than one log of clearance to be considered a valid treatment [4].

Factors to Consider

Many factors can influence the success of virus validation studies, and these should each be considered before embarking on a virus validation programme. The most important of these factors are outlined below.

Physical characteristics of virus

i) *Presence of an outer envelope*

As already mentioned, enveloped viruses are generally easier to inactivate and/or remove than their non-enveloped counterparts. Non-enveloped viruses are remarkably resistant to a variety of physiochemical treatments. However, non-enveloped virus should be included in the final virus selected for validation studies to demonstrate the broad applicability of the virus clearance regimen.

ii) *Size and shape of virus*

Size and shape of the virus is of greater concern for removal by filtration or chromatography. Both factors must be considered before selecting an appropriate filter or chromatography method.

iii) *Purity of virus preparation*

The virus preparation should be free of cellular debris or particulate, as this may over- or under-estimate the level of clearance. Pre-filtration of the solution before initiating the study can minimize this [4].

iv) *Extent of aggregation*

Viruses are prone to aggregation, and this is an important factor to consider in relation to virus clearance, either by filtration or inactivation. Viral aggregates may lower the efficiency of virus inactivation, as famously demonstrated in 1955, when formalin treatment of live polio vaccine failed to inactivate the poliovirus preparation completely. The occurrence of aggregation should be minimized during virus clearance trials to prevent over- or under-estimation of clearance [2].

v) *Reproducibility of virus behaviour*

In accordance with good laboratory practice (GLP), the testing should be reproducible. The use of well-characterised viral strains is of particular importance for inactivation studies to ensure reproducibility between tests.

vi) *Ability to culture virus to high levels, and relevance to manufacturing process*

The virus used for clearance studies will depend largely on the raw materials used in the product. The viruses of greatest concern for safety are indicated in Table 1. However, clearance testing with some of those listed (e.g. Hepatitis B and C) is not feasible due to insufficient yields, inadequate assay sensitivity or prolonged testing times [10]. Therefore, another virus sharing similar characteristics is used (e.g. use of murine leukemia virus (MuLV) as a model for non-infectious endogenous retroviruses) [4].

However, the chosen virus should resemble viruses that may contaminate the product and demonstrate a broad range of physiochemical properties to represent a "worst-case scenario" [1-3]. Typically, in Phase II and III trials, up to five viruses are tested during the clearance studies, all of which should exhibit a broad range of characteristics to show the robustness of the process.

vii) *Method of detecting virus*

Generally, infectivity assays are required for use in clearance studies. These include plaque assays and assays based on tissue culture infectious dose ($TCID_{50}$). Plaque assays are performed using an overlay technique, whereby the sample containing the virus is mixed with a host-cell monolayer. After appropriate incubation, plaques are observed in the monolayer that indicates infectious viral particles. $TCID_{50}$ assays determine the reciprocal of the highest dilution that causes cytopathic effects (CPE) of the culture cells.

The advent of the polymerase chain reaction (PCR) has allowed for the detection of very low levels of virus, but it does not indicate if the virus detected is infectious or not. Therefore, while this could be used to evaluate the extent of virus clearance, additional studies showing equivalence between PCR and infectivity-based assays would probably be required [1].

Table 1: Viruses of concern to safety.

Product type	Virus	References
Biotechnology-based	Herpes simplex virus Non-infectious endogenous retrovirus Encephalomyocarditis virus	4
Plasma based	Hepatitis B and C Human Immunodeficiency Viruses (types 1 and 2) Human parvovirus B19 Hepatitis A virus Parvovirus	3
Biopharmaceutical-based	Murine leukemia virus Parvovirus Poliovirus 4	4

viii) *Level of clearance required*

A thorough understanding of the manufacturing process is required to determine the virus to test, and the level of clearance required. For example, with plasma derivatives, the presence of clinically significant viruses such as HBC and HCV is prohibited, and detection of such viruses will result in quarantine of the product [2-4]. Conversely, for biotechnology-derived products, the presence of endogenous retrovirus particles in rodent cell lines is allowed, even at a concentration of 10^6-10^9 particles/mL. This difference is based on a "real" risk versus a "theoretical" risk [4]. However, it is important to realize that failure to detect a specific virus does not necessarily indicate its absence from the product, and may be linked to insufficiently sensitive virus assays.

Properties of product

i) *Size of protein*

If the product contains large proteins (e.g. Factor VIII), filtration may prove unfavourable due to loss in recovery of the product [4,9]. Conversely, virus removal filters used for proteins in the size range 100-300 kD may allow the passage of small viruses with these proteins. Only by conducting preliminary feasibility studies using the actual product can the overall performance of the filter be established.

ii) *Protein concentration*

High protein concentrations may protect viruses from some inactivation mechanisms such as pH treatment, thereby constituting worst-case conditions [4]. Again, preliminary studies should be performed to determine the testing conditions for the final validation study.

Comparison of Mammalian Viruses to Bacteriophages

Since the introduction of virus to the manufacturing process is prohibited in accordance with good manufacturing practice (GMP) requirements, virus clearance studies are typically performed on a scaled-down version of the process [1,3]. For inactivation mechanisms, the choice of virus is critical, and the effectiveness of the process will be dictated by the physical properties of both the virus and the product. In contrast, for virus removal by size exclusion filtration, the size and shape of the virus are the dominant factors, though as already mentioned, consideration must be given to the product characteristics.

Both mammalian viruses and bacteriophages display many common physical characteristics, such as a similar range in size (approximately 25-300 nm) and shape (e.g. icosahedral), and contain either an RNA or DNA genome. Therefore, for filtration studies where virus removal is based on size exclusion, similar levels of reduction would be expected for both mammalian viruses and bacteriophages of similar size and morphology. Table 2 shows a range of mammalian viruses commonly used in virus clearance studies, with a list of similarly sized bacteriophages.

Table 2: Comparison of mammalian virus and bacteriophage sizes.

Mammalian virus	Size (nm)	Bacteriophage	Size (nm)
PPV CPV MVM B19 Polio-1	18-24 18-24 18-26 18-26 25-30	PP7 MS2	25 26
EMCV HAV HEV	25-30 27-32 27-34	ϕX174	25-30
BVDV WNV	40-70 40-60	PR772	53-78
ADV MuLV HIV-1 or 2	70-90 80-110 80-120	ϕ6 T1	75-86 50 x 100
BHV PRV EBV	120-300 120-200 120-300	T4	200

Numerous studies have reported virus titre reductions (defined as the $\log_{10}$ of the ratio of the total concentration of virus in the feed to that in the effluent) from filtration experiments, and some of these results are depicted in Tables 3-6. Comparison of the titre reductions of the bacteriophages PP7 and the mammalian virus Porcine Parvovirus (PPV) (Tables 3 and 4) show similar levels of clearance under various conditions. Similarly, the titre reductions for poliovirus (Sabin 2) and PP7 (Table 6) are also in good agreement with each other, irrespective of the molecular weight cut-off (MWCO) or composition of the filter.

However, it can be observed that the carrier fluid used to suspend the virus does influence the extent of clearance. The data in Table 6 indicate that higher titre levels were achieved for both virus types when suspended in ultrapure water (UPW). This suggests that adsorption of the virus (both to the filter and to other viral particles) was a secondary removal method. The lower titre reductions observed when Dulbecco's modified Eagle minimum essential medium (DMEM-10) supplemented with 10% foetal bovine serum was used as carrier fluid support this suggestion, as adsorption would be minimized under these conditions.

Table 3: Log titre reductions for the bacteriophage PP7 and porcine parvovirus (PPV) after filtering through a 47 mm Ultipor VF grade DV20 membrane [Unpublished data, Pall Corporation].

Challenge virus	Log titre reduction		
	10 mL	**30 mL**	**50 mL**
PPV	4.3	4.3	3.2
PP7	5.1	6.5	4.5

Table 4: Comparison of log titre reductions for the bacteriophage PP7 and porcine parvovirus after filtering through a 47 mm Ultipor VF grade DV20 membrane [Unpublished data, Pall Corporation].

Carrier fluid	Virus	Log titre reduction		
		Volume passed through filter (mL)		
		10	**30**	**50**
PBS+1 mg/mL IgG	PPV	8.2	8.2	8.5
	PP7	5.4	5.6	5.3
PBS+5 mg/mL IgG	PPV	6.6	6.1	5.5
	PP7	4.9	4.9	4.6
DMEM	PV	6.2	5.8	5.6
	PP7	6.4	4.9	4.3

DMEM Dulbecco's modified Eagle minimum essential medium

Table 5: Comparison of titre reductions between bacteriophages and mammalian viruses for virus clearance studies based on size-exclusion filtration. Adapted from Aranha-Creado & Brandwein [11] and Oshima et al [9].

Filter type	Mammalian viruses used and size (nm)	Log titre reduction	Bacteriophages used and Size (nm)	Log titre reduction
PVDF, Ultipor® VF grade DV50	MuLV (80-120)	>6.7	ϕ6 (75)	>8.7
	Poliovirus (25-30)	>2.3	PP7 (25)	<1
	SV40 (45)	5.7->6.1	PR772 (53)	>6.7-7.9
PES, Omega® 300K VR	MuLV (80-120)	4.9	ϕ6 (75)	4.7
PVDF, Viresolve™/ 70™	MuLV (80-120)	>6.7	ϕ6 (75)	>6.8
	Poliovirus (25-30)	3.5	ϕX-174 (25-27)	2.9

PVDF Polyvinylidene fluoride
PES Polyethersulfone

It is an absolute requirement to demonstrate the scale-up of the small-scale virus clearance study to the industrial scale process (e.g. flow rates, effective filtration areas). Table 7 demonstrates the equivalency of the titre reductions for PP7 between small-scale (47 mm membrane discs) and large- scale (10" cartridges) challenges when all other experimental conditions were kept identical.

These reports suggest that when clearance is based on size exclusion filtration, bacteriophages appear as a suitable model for mammalian viruses of similar size. Indeed, use of bacteriophages as surrogates for mammalian virus have been widely accepted in medical and environmental applications, and although this is outside the scope of this short review, further information is available in Aranha-Creado and Brandwein [11].

Table 6: Comparison of log titre reductions for the bacteriophage PP7 and Poliovirus (Sabin 2). Adapted from Oshima et al [10].

Carrier fluid	Virus	50,000 MWCO PAN	13,000 MWCO PAN	6000 MWCO PAN	6000 MWCO PS
UPW	PP7	7.8	7.8	>7.8	ND
	Polio	ND	>6.3	>6.6	>6.4
0.85% saline + 1% TSB	PP7	5.0	7.5	>7.7	ND
	Polio	ND	6.6	ND	ND
DMEM-10	PP7	4.3	7.7	7.6	7.4
	Polio	4.6	>6.5	ND	>6.4

PAN Polyacrylonitrile
PS Polysulfone
DMEM-10 Dulbecco's modified Eagle minimum essential medium with 10% foetal calf serum

Table 7: Comparison of log titre reduction values for the bacteriophage PP7 using Ultipor® VF grade DV20 membrane (both 47 mm membrane discs and 10" cartridges) [Unpublished data, Pall Corporation].

Trial number	Log titre reduction for PP7	
	47 mm disc	10" cartridge
1	3.51	3.76
2	3.99	3.48
3	4.58	4.21

Advantages of Bacteriophages for Virus Clearance Studies

Table 8 compares various parameters common to both bacteriophage and mammalian virus clearance studies. Perhaps the most obvious advantage of using bacteriophages in preliminary virus validation studies is the ease with which these challenges can be performed. Bacteriophages are considerably less hazardous, and can be used in a general laboratory setting with minimum specialised containment equipment required (the most commonly used bacteriophages (and their associated bacterial hosts) in clearance studies are Biohazard Level 2 or less).

The assays to enumerate bacteriophages are also rapid, with results generally obtained within 12-24 hours. Furthermore, high levels of bacteriophages can be attained (10^{9-12} pfu/mL) which allows for greater accuracy in enumerating the virus, and also permits higher virus clearance claims. This allows manufacturers an inexpensive means to optimize their clearance process, and predict clearance values before performing final testing with the chosen mammalian viruses. The broad size distribution of bacteriophages indicates that they can be used for preliminary studies irrespective of the final mammalian virus to be tested.

Table 8: Comparison of virus clearance studies using mammalian viruses and bacteriophages.

	Mammalian viruses	**Bacteriophages**
Assay sensitivity	Difficult to detect low numbers	Can detect ≤1 pfu/mL
Cost of challenge	High	Low
Time for results	> 1 week (typically)	12-24 hours
Skill level	High	Moderate
Titre levels	Low (10^{5-6} pfu/mL typical)	High (10^{9-12} typical)
Specialized equipment	Yes	Minimum

Conclusions

Based on reported data , it appears that bacteriophages can be used as a surrogate for mammalian virus when removal is based on size exclusion, and is independent of surface or genomic characteristics of the virus.

Although the FDA still requires virus clearance studies to be performed on mammalian viruses, studies using bacteriophages are a useful tool for preliminary clearance evaluations and process optimization. In addition, the use of bacteriophages is useful when establishing correlation of integrity test values to the removal properties of virus retaining filters.

Installation and validation of a virus clearance step in a manufacturing process is a complex procedure, and many factors must be considered to ensure successful implementation. A clear understanding of these factors, and preliminary studies using bacteriophages, will assist the user to have realistic expectations when size exclusion-based mechanisms, such as filtration, are used to effect viral clearance.

References

1 Rev 2, Note for guidance on virus validation studies: the design, contribution and interpretation of studies validating the inactivation and removal of viruses. CPMP/BWP/268/95.

2 ICH Harmonized Tripartite Guideline. Viral safety evaluation of biotechnology products derived from cell lines of human or animal origin, 1997.

3 Guide to Inspections of viral clearance processes for plasma derivatives. http:www.fda.gov/ora/inspect_ref/igs/viralcl.htm.

4 Aranha H: Viral clearance strategies for biopharmaceutical safety. Part 1-3. Biopharm 2001. http:www.pall.com/biopharm/pdf/bp5560.pdf and http:www.pall.com/biopharm/pdf/bp5561.pdf.

5 Brorson K, Krejci S, Lee K, Hamilton E, Stein K, Xu Y: Bracketed generic inactivation of rodent retroviruses by low pH treatment for monoclonal antibodies and recombinant proteins. Biotech Bioeng 2003;82:321-329.

6 Validation guide for Pall Ultipor® VF Grade DV20 AB style virus removal filter cartridges, VG-DV20.

7 Validation guide for Pall Ultipor® VF™ DV50 Ultipleat™ AB style virus removal filter cartridges. TR-UDV50.

8 Aranha-Creado H, Oshima K, Jafari S, Howard G Jr, Brandwein H: Virus retention by a hydrophilic triple-layer PVDF microporous membrane filter. PDA J Pharm Sci Tech 1997;51:119-124.

9 Oshima KH, Evans-Strickfaden TT, Highsmith AK: Comparison of filtration properties of Hepatitis B virus, Hepatitis C virus and Simian virus 40 using a polyvinylidene fluoride membrane filter. Vox Sanguis 1998;75:181-188.

10 Oshima HK, Evans-Strickfaden TT, Highsmith AK, Ades EW: The removal of phages T1 and PP7, and poliovirus from fluids with hollow-fiber ultrafilters with molecular weight cur offs of 50000, 13000, and 6000. Can J Microbiol 1995;41:316-322.

11 Aranha-Creado H, Brandwein H: Application of bacteriophages as surrogates for mammalian viruses: a case for use in filter validation based on precedents and current practices in medical and environmental virology. PDA J Pharm Sci Tech 1999;53:75-82

Dr. M. McAlister, SLS, Pall Corporation, 25 Harbor Park Drive, Port Washington, NY 11050, U.S.A.
E-mail: morven_mcalister@pall.com

Robertson J (ed): PDA/EMEA European Virus Safety Forum.
Dev Biol. Basel, Karger, 2004, vol 118, p 99.

SESSION III (2)

Virus Validation, Model Viruses

Chairs: *H. Willkommen* (Clearant)
G. Rössling (Schering)

Robertson J (ed): PDA/EMEA European Virus Safety Forum.
Dev Biol. Basel, Karger, 2004, vol 118, pp 101-105.

West Nile Virus: Recent Experience with the Model Virus Approach

T.R. Kreil

Global Pathogen Safety, Baxter BioScience, Vienna, Austria

Key words: Blood transfusion, blood-borne pathogens, transmission, virus inactivation, emerging virus, model virus.

Abstract: During recent West Nile Virus (WNV) epidemics in the U.S.A., virus transmissions through transfusion of blood components have been observed, also raising concerns about the safety of plasma derivatives. To verify the safety margins of these products, as initially shown using model viruses very similar to WNV, virus inactivation procedures incorporated into their manufacturing processes were re-investigated using WNV directly.
The results show that WNV behaves exactly as had been predicted based on available data for similar ("model") viruses, i.e. it is readily inactivated by e.g. pasteurisation of human albumin, solvent-detergent treatment of intravenous immunoglobulin and factor VIII, vapour heating of factor VIII inhibitor by-passing activity, and incubation of intravenous immunoglobulin at low pH. Our investigation thus verifies the safety margins of plasma derivatives against Flaviviruses in general, and now WNV in particular. Also, use of the model virus concept was confirmed to be well capable of predicting the behaviour of a certain virus based on available data for closely related viruses.

INTRODUCTION

While throughout history a number of viruses have challenged the safety of biological medicinal products, not all these viruses can be directly studied using practical infectivity assays. Instead, taxonomically related viruses with equivalent physicochemical characteristics, so called model viruses, have been and continue to be experimentally used to understand the behaviour of these other relevant viruses during the manufacturing processes of plasma derivatives and other biotechnology products. Particularily for two major transfusion-transmitted viruses, i.e. the hepatitis viruses B (HBV) and C (HCV), an infectivity assay is still unavailable, and consequently all available information about product safety versus any potential transmission of these two viruses has been generated using the model virus concept.

Moreover, for parvovirus B19 (B19), a virus more recently recognized as being of a certain relevance, a practical infectivity assay is not available, although some progress is being made in establishing B19 infectivity assays [1].

The model virus concept and many aspects relevant for its reduction to practice is detailed in regulatory guidance on the conduct of virus reduction studies [2]. Key elements of the guidance relative to the choice of model viruses (author's selection, emphasis added) are:

4.1 "Viruses for validation studies should be chosen firstly to resemble viruses which could contaminate the product as closely as possible and, secondly, to represent as wide a range of physico-chemical properties as possible to ***test the ability of the system to eliminate viruses in general***."

4.2 "Most validation studies employ laboratory strains of virus which can be produced and assayed conveniently. ... different laboratory strains of virus may have different properties from each other and from naturally occurring viruses. Consequently, ***any virus used in a validation study is actually a model virus***."

4.3 "Examples of the choice of viruses are: (i) Human plasma-derived clotting factors concentrates ***have been contaminated*** by HIV. ***Thus*** the production process for such materials ***must be evaluated*** for its ability to inactivate/remove infectious HIV."

In conclusion, model viruses are used for understanding the ability of any manufacturing process to deal with any potential contamination, and their use is an inherent characteristic of virus reduction studies. Beyond this general approach only a specifically identified and medically significant risk requires further consideration, i.e. the appropriate response is determined by the level of risk (risk-based approach).

West Nile virus (WNV) is a member of the *Flaviviridae* family of viruses, genus *flavivirus*, approx. 50 nm in diameter with a lipid-envelope and an approx. 10.5 kilobase positive-strand RNA genome [3]. The virus was first isolated in 1937 from the blood of a febrile woman in the West Nile district of Uganda. During 1957 it was recognized as the cause of severe meningo-encephalitis in elderly patients during an outbreak in Israel. In 1996 it caused the first (recognized) outbreak in Europe, with approximately 450 cases in Romania, and then in 1999 it was firstly detected on the East coast of the U.S.A., causing approximately 60 clinical cases of WNV infection [4]. While the appearance of WNV in New York received significant media attention, rationally it should have been of little surprise, given the global occurrence and migratory activity of *flaviviruses* in general. Already after this first occurrence of WNV in the USA, predictions – and as we now know very reasonable ones – about a potential transfusion transmission were made [5].

WNV established its presence in the U.S.A. during the following years, and then caused epidemics of unprecedented size during the 2002 and 2003 mosquito seasons, which involved 4156 and 6507 reported cases of WNV infection, unfortunate 284 and 136 WNV-attributed fatalities (2003 numbers are as at October 10, 2003 [6]). Beyond the individual losses, however, during the 2002 season WNV transmission through labile blood products for direct transfusion was documented [7], as earlier predicted. In total, WNV transmission to 23 recipients was confirmed through red blood cell cells, thrombocytes and fresh frozen plasma obtained from

16 individual donations [8]. This challenged the safety of the blood supply, a system which provides labile blood products for transfusion to the entire U.S. population.

In guidance issued by the competent authority for the safety of the blood supply in the USA, the Food and Drug Administration [9], donor deferral criteria relative to WNV were specified to enhance the safety of labile blood components for transfusion. For stable blood products, i.e. plasma derivatives, the conclusion was that "FDA has reviewed the viral reduction processes in place for all plasma derivatives. The methods in place have been validated to inactivate flaviviruses related to WNV." Consistent with the model virus concept, information obtained by using viruses related to a particular virus of concern formed the basis for an evaluation of the level of risk from an agent for which no direct information was available.

From an earlier application of the model virus concept, i.e. modelling the behaviour of HCV for which an assay is not available by using physicochemically equivalent viruses, a very substantial body of information was already available for other *Flaviviruses*, particularly for Bovine Viral Diarrhoea Virus (BVDV) and Tick-borne Encephalitis virus (TBEV). From studies with BVDV and TBEV it was well established that the lipid-enveloped *Flaviviruses* are all susceptible, as expected, against solvent-detergent or intensive detergent treatments, and also to heat treatments such as pasteurisation and vapour heating.

While scientifically it was felt that WNV was quite well understood and did not pose any real threat to the safety of plasma derivatives, the concerns of patients who use these products were taken seriously by several partners in the biotechnology industry.

Materials and Methods

A WNV specimen, obtained from the liver of a Snowy owl found dead in New York during the first WNV outbreak in 1999 (isolated by Dr. Tracy MacSamara, kindly provided by Dr. Robert E. Shoppe; isolated by filtration of the liver, passage once on Vero cells, and lyophilised: isolate 385-99; partially characterized by genome sequencing; [10]) was consequently obtained and an infectivity assay established for the low-speed centrifugation clarified supernatant of its third serum-free Vero cell passage. This tool was then used to investigate WNV inactivation during several of the most commonly used inactivation methods such as solvent-detergent, pasteurisation, vapour heating and low pH treatment, in direct comparison to the behaviour of the earlier used model viruses. For the purpose, down-scaled versions of these processes were established, and the equivalence with their respective large scale counterparts validated by determining a variety of process and product parameters for comparison of both processes. For parameters critical for virus inactivation, e.g. the concentration of solvent-detergent chemicals and the temperature and incubation times for heating processes, conditions at or just below the manufacturing process specifications, were chosen to reflect a situation least favourable for virus inactivation (worst case).

Results

The results from this verification study confirmed that all the *Flaviviruses* tested, i.e. BVDV, TBEV and now WNV, behave very similarly, i.e. they are all very sensitive to the inactivation methods investigated. The study thus supports the validity of the model virus concept, and the safety of plasma products in general (for further detail see: [11]).

Discussion and Conclusion

The European Committee for Proprietary Medicinal Products (CPMP), after being presented with the results of this and similar studies, issued a Position Statement on West Nile Virus and Plasma-derived Medicinal Products saying that "... the data presented ... provided reassurance that the steps currently in place are adequate to assure viral safety of plasma-derived medicinal products." [12].

Additional factors for consideration within a risk-based approach would include a comparison of the potential disease outcome for WNV versus e.g. the "must be evaluated" [2] HIV, i.e. an approx. 1:150 chance of severe diseases versus an inevitably fatal infection. Also, where for a chronic virus infection, e.g. HIV, testing to defer a potential donor permanently is indeed desirable, a WNV-infected donor would, after a viraemic period of only a few days, contribute WNV antibodies to a manufacturing plasma pool, and would thus actually βenhance the safety margins of products from these pools.

Acknowledgements

The principal investigators of this study, Johanna Kindermann and Andreas Berting, are complemented for their thorough work under extremely pressing deadlines. We are indebt to Otfried Kistner for providing West Nile virus and assay development details. The contributions of the entire Pathogen Safety team, most notably Alexandra Danzinger, Karin Berka, Bettina York and Kerstin Holzinger (cell culture, virus propagation), Friedrich Schiller and Nicole Wurzer (SD, pasteurisation, and low pH), Ulrike Unger and Lissy List (vapour heating), Dr. Geza Szabo, Elke Schabhuettl and Astrid Beckmann (data monitoring and compilation) are gratefully acknowledged.

References

1 Blumel J, Schmidt I, Willkommen H, Lower J: Inactivation of parvovirus B19 during pasteurization of human serum albumin. Transfusion 2002;42:1011-1018.

2 CPMP/BWP: Note for Guidance on Virus Validation Studies: The Design, Contribution and Interpretation of Studies validating the Inactivation and Removal of Viruses. CPMP/BWP/268/95 adopted February 1996.

3 Heinz FX, Collett MS, Purcell RH, Gould EA, Howard CR, Houghton M, et al: Family Flaviviridae, in Van Regenmortel MHV, Fauquet CM, Bishop DHL, Carstens EB, Estes MK, Lemon SM, et al (eds): Virus taxonomy: classification and nomenclature of viruses. 7th Report of the International Committee on Taxonomy of Viruses. San Diego, Academic Press, 1999, pp 859-878.

4 Petersen LR, Roehrig JT: West Nile virus: a re-emerging global pathogen. EID 2001;7:611-614.

5 Biggerstaff BJ, Petersen LR: Estimated risk of West Nile virus transmission through blood transfusion during an epidemic in Queens, New York City. Transfusion 2002;42:1019-1026.

6 CDC ArboNET: West Nile Virus Activity – United States, October 2-8, 2003. MMWR 2003;52:1022-1023.

7 CDC: Investigation of blood transfusion recipients with West Nile virus infections. MMWR 2002;51:823.

8 Pealer LN, Marfin AA, Petersen LR, Lanciotti RS, Page PL, Stramer SL, et al: Transmission of West Nile virus through blood transfusion in the United States in 2002. N Engl J Med 2003;49:1236-1245.

9 FDA: Final Guidance for Industry, October 25, 2002. Recommendations for the Assessment of Donor Suitability and Blood and Blood Product Safety in Cases of Known or Suspected West Nile Virus Infection. (revised May 2003: http://www.fda.gov/cber/gdlns/wnvguid.pdf).

10 Beasley DW, Li L, Suderman MT, Barrett AD: Mouse neuroinvasive phenotype of West Nile virus strains varies depending upon virus genotype. Virology 2002;296:17-23.

11 Kreil TR, Berting A, Kistner O, Kindermann J: West Nile virus and the safety of plasma derivatives: verification of high safety margins, and the validity of predictions based on model virus data. Transfusion 2003;43:1023-1028.

12 EMEA/CPMP/BWP: CPMP Position Statement on West Nile Virus and Plasma-derived Medicinal Products (EMEA/CPMP/BWP/3752/03/Adopted, September 2003).

Dr. T.R. Kreil, Global Pathogen Safety, Baxter BioScience, Benatzkygasse 2-6, A-1220 Vienna, Austria.
E-mail: thomas_kreil@baxter.com

Robertson J (ed): PDA/EMEA European Virus Safety Forum.
Dev Biol. Basel, Karger, 2004, vol 118, pp 107-112.

Cell Culture-Based Assay of Parvovirus B19 and the Relevance of Animal Model Viruses

J. Blümel

Paul-Ehrlich-Institut, Langen, Germany

Key words: Parvovirus B19, infectivity assay, virus validation, model virus

Abstract: One of the most challenging problems with the interpretation of virus validation studies is the relevance of the model viruses investigated. There are cases where model viruses have been shown to predict the inactivation/removal of a specific virus while unexpected results were obtained in other cases such as inactivation of parvovirus B19 by pasteurisation or treatment with guanidinium hydrochloride. The understanding of the mechanism underlying virus inactivation/removal might be helpful for the selection of model viruses and for the assessment of validation data on its relevance for a specific virus in question.

INTRODUCTION

Virus validation is a model approach. In general, virus is spiked into the process material from production and the spiked material is processed in the laboratory in a down-scaled version of the manufacturing step to be investigated. The virus reduction factor is determined by the difference of virus load in the material before and after processing. The choice of viruses which have been used in such studies is of outstanding importance for the interpretation of the results. Two aspects have to be considered when selecting viruses for validation studies. Viruses for validation should first represent or resemble viruses which may contaminate the product and second represent as wide a range of physico-chemical properties as possible, to test the ability of the system to eliminate viruses in general [1]. In many cases, it is not possible to use the 'relevant virus species' which may contaminate a product, since suitable cell culture systems for measuring infectious virus are not available. For example, human immunodeficiency virus (HIV) and hepatitis A virus (HAV) may contaminate human plasma and can be used for the validation of human plasma-derived products, while animal model viruses have to be used for Hepatitis C virus (HCV). The use of Hepatitis B virus or a suitable animal hepadna virus is still under discussion. Parvovirus B19 (B19V) is also a virus which may contaminate human plasma at high levels and cannot be readily propagated in cell culture. In this study we focused on the development of a laboratory assay for B19V in order to compare the inactivation of B19V with animal model viruses. B19V inactivation differed strikingly from animal parvovirus inactivation. These data and others on 'relevant' and model viruses will be discussed regarding the general aspects to be considered for the choice of viruses for validation studies and the interpretation of such results.

Cell Culture-Based Assay of Parvovirus B19

Human Parvovirus B19 (B19) is a frequent contaminant in plasma donations. Viraemic blood donors are mostly asymptomatic and generally not excluded by screening for B19 DNA or B19-specific antigen. The frequency of B19 DNA-positive blood donations has been found to vary between 1:800 and 1:1590. Viraemia may be up to 10^{12} genome equivalents (geq) per mL, albeit at a lower frequency of 1:8000 [2]. This can lead to a high contamination of large plasma pools and current purification procedures are not sufficient to remove the high B19 loads from the final products completely. Parvoviruses are small non-enveloped viruses not affected by inactivation procedures which are highly effective against enveloped viruses (e.g. solvent/detergent-treatment). Furthermore, the effectiveness of other inactivation procedures (e.g. heating steps) against B19 remains unclear. Due to the lack of a suitable in vitro cultivation system for B19V, animal parvoviruses have to be used as a model for B19V in validation studies, but it remains uncertain whether these models accurately reflect inactivation of B19. We used KU812Ep6 cells [3] to establish a system which is suitable for investigation of B19V inactivation (Fig. 1).

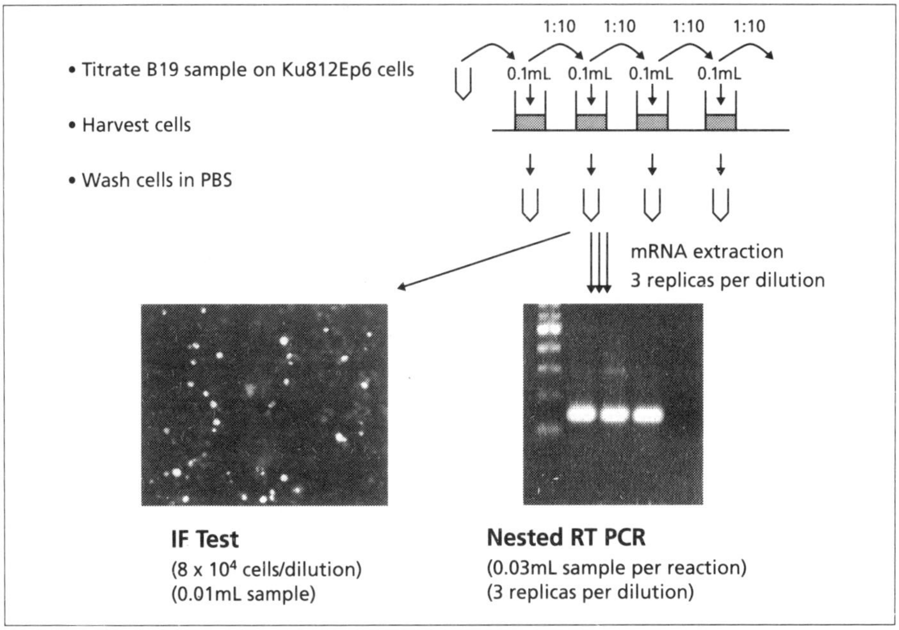

Fig. 1: Overview of the cell culture-based assay for infectious B19V.

The infectious B19V assay is performed by Immune Fluorescence Microscopy (IF) and nested Polymerase Chain Reaction (PCR). Only 5 to 10% of cells stain with anti-capsid antibody, indicating that KU812Ep6 is not a homogeneous cell line. Nevertheless, a proportional relation between virus titre and the number of infected cells could be demonstrated. The primer design for the nested PCR allowed differentiation between mRNA and DNA which is particularly important in view of the high titres of the virus inoculum. Virus detection by mRNA was more sensitive than the IF assay: 1 IFU = 5.6×10^4 IU B19 DNA, $1 mRNA_{50} = 4.3 \times 10^3$ IU B19V DNA. Titration of various IgG and IgM negative Parvovirus B19 positive samples confirmed this relation.

Parvovirus B19 inactivation on pasteurisation of albumin [4].

Parvovirus B19 inactivation by pasteurisation was measured at different temperatures and with different albumin concentrations (5, 20, 25%) and preparations. It has been previously demonstrated that animal parvoviruses are not inactivated at 60°C. This previous observation was confirmed by pasteurizing 5% albumin at 60°C where PPV remained very stable over a period of 60 min. However, parvovirus B19 was rapidly inactivated under the same conditions (Fig. 2). A different property of the virus capsid could be the reason for the different thermosensitivity of Parvovirus B19 and animal parvoviruses. At lower temperatures (55.5°C, 54°C) inactivation of Parvovirus B19 significantly declines and at 53°C no effective inactivation can be achieved. The mechanism of virus inactivation during pasteurisation was investigated by a DNAse I assay. When the virus capsid is heavily damaged or

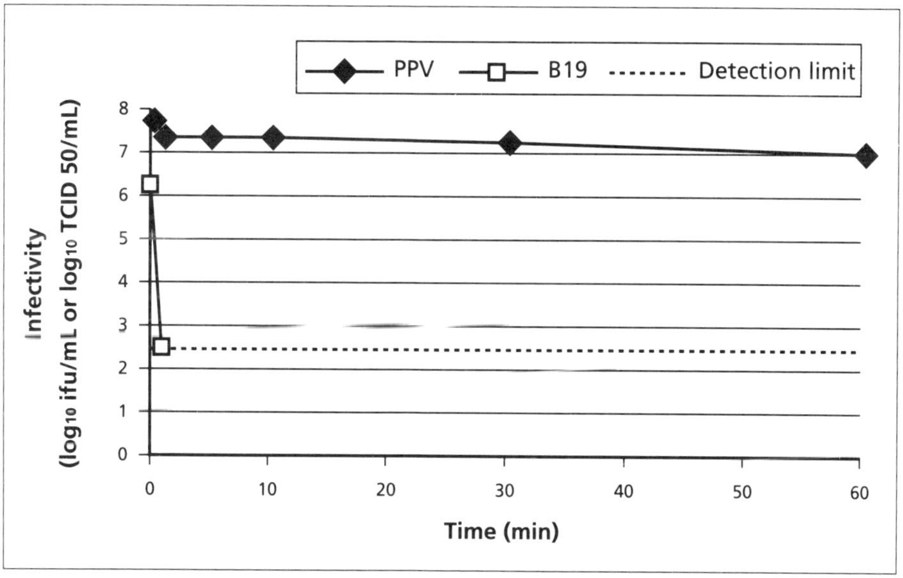

Fig. 2: **B19V and PPV during pasteurisation of albumin.** Albumin 5% was pre-heated to 60°C and spiked 1:11 with B19V-containing plasma or PPV. Incubation at 60°C was continued and infectious titre was determined after several time points by titration on KU812Ep6 and PK-13 cells. The starting point of inactivation kinetics (0min) represents the infectious titre from a sample spiked at room temperature.

disaggregates during heating, the virus DNA is no longer protected by the complete capsid and becomes exposed to DNAse I resulting in DNA degradation. The correlation between the grade of virus inactivation and DNAse I degradation indicated that destruction of virus capsids by heat is indeed the mechanism that causes virus inactivation.

Virus inactivation in guanidinium hydrochloride

Guanidinium hydrochloride (Gu-HCl) can be used as buffer for the sanitisation of various columns such a protein A columns or immuno-affinity columns. The resin of such columns may be used many times (up to 200-fold or more) in the manufacture of biological products. To investigate whether there is a risk of virus accumulation on the resin during extended re-use, validation studies are performed which measure virus inactivation after spiking into the sanitisation buffer. We performed such studies using a broad variety of enveloped and non-enveloped viruses. A summary of the results of the study is shown in Table 1. B19V was susceptible to GuHCl at concentrations above1M, while animal parvoviruses were resistant at concentrations up to 6M. The pattern of inactivation was also not uniform for other non-enveloped viruses. Enteroviruses (Polio virus and HAV) were completely resistant while Adenovirus Type 5 (Ad5) and Feline Calicivirus (FCV) were inactivated even at low concentrations. Polyomavirus SV40 showed some intermediate stability. This study demonstrates that data from an inappropriately-selected panel of model viruses may be misleading when reaching conclusions on the inactivation capacity of viruses in general. For example, the selection of FCV, Ad5, and B19V

Table 1: Virus reduction factors (log_{10}) after 2h incubation of virus in 0.5 to 6M GuHCl, pH7.2 at room temperature.

Virus inactivation in GuHCl					
Virus	**0.5M**	**1M**	**2M**	**4M**	**6M**
PRV	1.0	3.1	≥2.3		
BVDV	1.1	1.9	≥4.6	≥3.9	≥3.9
SFV		2.9	3.4	=3.3	
VSV		0.9	3.6	≥2.9	
SV40	1.4	3.7	3.8	4.0	4.7
FCV		3.3	≥5.4	≥5.4	
AdV			≥4.8	≥4.8	≥4.8
MVM		(0.0)	(0.0)	(0.6)	(0.9)
B19	1.1	2.4	≥4.1		
PPV			(0.48)	(0.65)	1.1
HAV			(0.6)	(0.3)	1.7

from three different non-enveloped virus families might lead to the conclusion that 2M GuHCl inactivates non-enveloped viruses in general. However, this is clearly not the case for animal parvoviruses and enteroviruses. An explanation for the different susceptibility of non-enveloped viruses could be the stability of virus capsids. GuHCl is a chaotropic agent which inactivates viruses, presumably by the disintegration of virus capsids. It is known that the resistant viruses (animal parvoviruses, poliovirus and HAV) are transmitted via the oral-faecal route and probably pass through the stomach and duodenum. This requires stability of virus capsids against low or high pH as well as against chaotropic substances such as bile acid which is present in the duodenum. In contrast, adenoviruses, FCV, SV40, and B19V are not transmitted via the oral-faecal route and do not pass through the stomach and duodenum.

Discussion

The adequacy of the model virus is one of the most challenging problems when interpreting the results from virus validation studies. There are two aspects which have to be considered when selecting viruses for validation studies [1]. Viruses for validation should be selected firstly to resemble viruses which may contaminate the product and secondly to represent as wide a range of physico-chemical properties as possible, to test the ability of the system to eliminate viruses in general.

When choosing viruses for validation, it should be kept in mind that the aim of virus inactivation/removal steps is to cover a broad variety of viruses. Previous experience clearly shows that the source material may contain unknown viruses and that new viruses may appear. Even a thorough testing of blood, other biological starting material, or cell banks, is not sufficient to exclude every virus. The introduction of virus inactivating/removing steps is a complementary approach ensuring a much higher level of safety than can be achieved by testing measures alone. To provide evidence that a process step can inactivate/remove a broad variety of viruses, the spectrum of viruses should consider a broad variety of viruses. As it is not possible to predict the nature of unknown or emerging viruses, any validation data have to be considered as data from model viruses.

When the focus is on a specific virus which may contaminate a biological starting material or cell line, using the 'relevant virus' for validation can avoid misinterpretation from model viruses. During pasteurisation of albumin, hepatitis A virus is inactivated considerably more slowly than animal model enterovirus EMCV, and B19V is sensitive to pasteurisation in contrast to animal model parvovirus PPV. However, this does not mean that model viruses would cause inadequate results in each case. Comparable inactivation has been recently shown for West Nile Virus and enveloped animal pestiviruses and flaviviruses at certain key steps such as treatment with solvent/detergent or pasteurisation [5]. Virus inactivation at solvent/detergent treatment occurs by destruction of the virus envelope by the combined action of solvent and detergent. As the virus envelope from these viruses has a similar structure, comparable inactivation might have been expected. In contrast, vaccinia virus has a much more complicated double envelope and it was observed that this virus was less sensitive to S/D-treatment [6]. In conclusion, the mechanism of virus inactivation/removal determines whether model viruses are adequate or not. For example, non-enveloped viruses will not be used for validation of S/D-treatment as no inactivation occurs and validation of S/D-treatment should include viruses with different properties of the envelope to cover a broad range of enveloped viruses. In contrast, validation of filtration should consider the particle size of investigated viruses. Validation of UV irradiation or gamma irradiation should consider the properties of virus genomes as the viral genome is the primary target of inactivation. In summary, an understanding of the mechanism of virus Inactivation/removal of a specific process is clearly helpful for the interpretation of the relevance of validation data which are obtained by using model viruses.

References

1 CPMP Note for Guidance on Virus Validation Studies: the design, contribution and interpretation of studies validating the Inactivation and removal of viruses. CPMP/BWP/268/95. 1996.

2 Weimer T, Streichert S, Watson C et al: High-titer screening PCR: a successful strategy for reducing the parvovirus B19 load in plasma pools for fractionation. Transfusion 2001;41:1500-1504.

3 Miyagawa E, Yoshida T, Takahashi H et al: Infection of the erythroid cell line, KU812Ep6 with human parvovirus B19 and its application to titration of B19 infectivity. J Virol Methods 1999;83:45-54.

4 Blümel J, Schmidt I, Willkommen H, Löwer J: Inactivation of parvovirus B19 during pasteurisation of human serum albumin. Transfusion 2002;42:1011-1018.

5 Kreil TR, Berting A, Kistner O et al: West Nile virus and the safety of plasma derivatives: verification of high safety margins, and the validity of predictions based on model virus data. Transfusion 2003;43:1023-1028.

6 Roberts P: Resistance of vaccinia virus to inactivation by solvent/detergent treatment of blood products. Biologicals 2000;28:29-32.

Dr. J. Blümel, Paul-Ehrlich-Institut, Paul-Ehrlich-Straße 51-59, D-63225 Langen, Germany.
E-mail: blujo@pei.de

Robertson J (ed): PDA/EMEA European Virus Safety Forum.
Dev Biol. Basel, Karger, 2004, vol 118, pp 113-122.

Methods for Validation of Hepatitis B Virus Inactivation

W.H. Gerlich, D. Glebe

Institute for Medical Virology, National Consulting Laboratory for hepatitis B, C and Blood-Transmissible Virus Diseases, Justus-Liebig University, Giessen, Germany

Key words: HBV, DHBV, disinfection, Tupaia belangeri, realtime PCR

INTRODUCTION

Until the discovery of Australia antigen and its relationship to "serum hepatitis" in the late 1960s, hepatitis B virus (HBV) was a frequent and unavoidable contaminant of human biological materials such as blood, organs and tissues. With the introduction of sensitive tests for the surface antigen of HBV (HBsAg), detection and exclusion of donors who were infected with HBV became possible and has been common practice since the early 1970s. There are still certain gaps in the detection and exclusion of HBV positive donors: in spite of their low detection limit at 10-100 pg HBsAg/ml, current assays for HBsAg cannot completely exclude donors with fewer than 10^5 HBV particles/ml. In the theoretically worst case, 10^6 HBV particles/ml may be required to generate a positive result of a sensitive HBsAg assay. Such a situation occurs usually in the early phase of HBV infection even months before disease symptoms occur and antibodies against the HBV core antigen (anti-HBc) are formed [1,2]. A second problem is low-level carriers of HBV who are anti-HBc positive, HBsAg negative, but HBV DNA positive and potentially infectious [3,4]. Finally, HBV is a variable virus with high mutational capacity if there is immune or drug-induced selection. Escape mutants of HBsAg which are difficult to detect occur after vaccination of subjects who were already infected [5] or during reactivation of HBV under T cell suppression [6].

Due to these gaps, inactivation of HBV in biological materials is desirable. Furthermore, there are situations where testing for HBsAg is not feasible. Inactivation of HBV requires validation of the applied methods but this has turned out to be a difficult problem. Various endpoint titrations in chimpanzees as experimental animals have revealed titres of 50 % infectious doses (ID50) around 10^8/ml for typical highly viraemic HBV carriers containing HBsAg and the immunomodulator HBeAg [7]. In HBsAg-negative HBV infections, ID50 titres are not known but they must clearly be lower than the physical number of HBV particles. Usually this number will be < 10^6/ml but in the case of some HBsAg escape mutants, very high numbers around 10^9/ml are possible ([6] and W. Gerlich, unpublished observations). Thus, a high capacity of inactivation procedures and a wide quantitative range of infectivity testing are required for HBV. A problem closely related to HBV inactivation in biological materials is the validation of disinfectants for HBV activity.

Infectivity assays for HBV

HBV is a member of the hepadnaviridae family. Hepadnaviruses are highly species- and liver-specific. Furthermore, their replication depends strictly on the physiological state of differentation as it occurs in resting hepatocytes within the organ. While ID50 titres of HBV carrier sera in experimentally infected chimpanzees are very high [7,8], more available experimental animals such as rodents are not susceptible at all. HBV and the other hepadnaviruses require liver-specific transcription factors for their replication. Thus, some human and other cell lines of hepatic origin are permissive for HBV replication after artificial introduction of the HBV genome, but they are not susceptible to infection, i.e. physiological entry of HBV. Only primary hepatocyte cultures can be *infected* with hepadnaviruses. Until recently, freshly excised human or primate livers were required for generation of HBV-susceptible hepatocyte cultures [9]. But even such cultures show wide variations in susceptibility and low efficiency of replication [10]. They are not easily available and not suitable for reliable titrations of HBV infectivity.

A somewhat more available and feasible source of HBV susceptible liver tissue are Tupaias, small squirrel-like animals (Fig. 1) living in Southeast Asia. They belong to a single order, scandentia, which branched off from the other primates and rodents c. 90 million years ago. They still show some features related to primates,

Fig. 1: Tupaia belangeri can be kept in captivity but needs special climatic conditions since the natural habitat is the tropical forests in Southeast Asia (picture taken by Frauke Ohl, Deutsches Primatenzentrum, DPZ, Göttingen).

and one of these features is that their cultured hepatocytes are at least as susceptible to HBV than most primary hepatocyte cultures of human origin [11,12]. Tupaias can be kept and bred in captivity. Tupaias are used on a large scale for behavioural and pharmacological studies. Thus, Tupaias are readily available from centres (e.g. the Deutsches Primatenzentrum Göttingen) where these animals are bred. One Tupaia liver usually yields 50-100 cultures of c. 10^5 hepatocytes in 2 cm-dishes (Fig. 2A, B). The cultures bind and take up HBV specifically (Fig. 2C, D), but produce relatively little progeny virus. No cytopathic effect is visible. As a marker of infection by viable virus, expression of viral products is measured. Most sensitive is RT/PCR of HBV mRNAs. Less demanding is the detection of the secreted HBV antigens, HBeAg or HBsAg, 9-12 days post infection, using commercially available ELISAs (Fig. 3A). The detection limit is currently one HBV particle per hepatocyte and 10^5 hepatocytes/culture (Fig. 3B). A peculiarity of the system is that it requires purified HBV particles from HBV carrier serum or plasma (Fig. 4). The reason is that human serum contains an inhibitor of HBV infectivity which can be removed by sedimentation of the virus in a density gradient [11]. The system has already been successfully used for the assay of HBV neutralising antibodies ([12] and unpublished results) and for the development of antiviral peptides suppressing infectivity of HBV (Glebe et al, in preparation). The system may be suitable for validation of inactivation procedures for HBV, but its specific requirements limit its general applicability. In particular, purified virus has to be used at a rather high input of c. 10^9 particles/ml to demonstrate an inactivation factor of $\geq 10^4$.

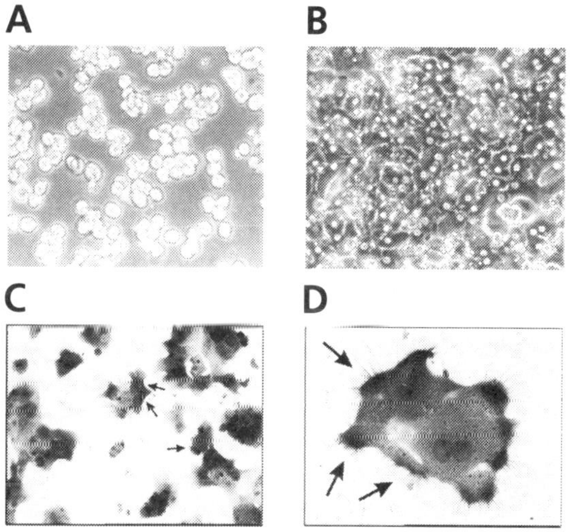

Fig. 2: (A, B) Morphology of primary Tupaia hepatocyte cultures. Phase contrast micrographs of suspended primary hepatocytes after isolation (A) and after attachment to collagen-coated culture plates (B), (magnification x100). (C, D) Binding and uptake of subviral particles to primary Tupaia hepatocytes. Immune staining of Tupaia hepatocytes after incubation with purified HBV subviral particles for 1h at 37°C against the S domain (black, arrows) (C), (magnification x 100). HBs particles (arrows) are located at distinct areas of the plasma membrane (D), (magnification, x400).

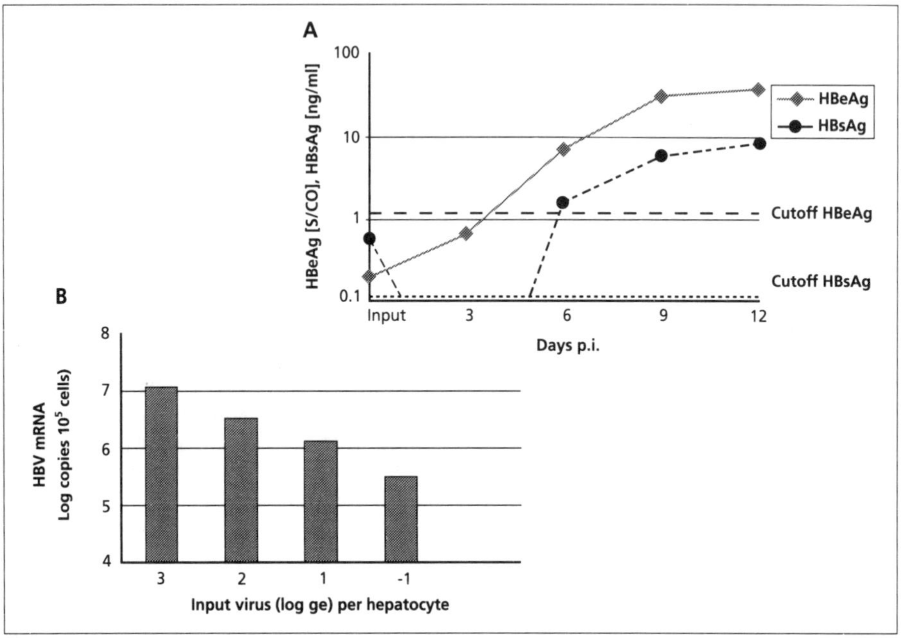

Fig. 3: (A) HBeAg and HBsAg secretion of primary Tupaia hepatocyte cultures infected with 100 genome equivalents per hepatocyte (ge/cell) in 3 cm dishes (10^5 hepatocytes). Medium was changed every three days, so each point represents HBeAg or HBsAg in the supernatant newly synthesized within three days. S/CO, sample to cut-off signal. (B) Detection of total HBV mRNA pool in cytoplasmic extracts from Tupaia hepatocytes 12 days after infection with increasing amounts of purified HBV (real-time RT-PCR of X-region, LightCycler System). ge, genomic equivalents.

Recently a hepatoma cell line, Hepa RG, has been described which can, in contrast to primary hepatocytes, be propagated in vitro. After re-differentiation of the cells in a non-dividing state with dimethylsulphoxide (DMSO) for three weeks, the cells become susceptible to HBV. This cell line has been useful for detection of neutralising antibodies [13] and antiviral peptides (P. Gripon and S. Urban, manuscript submitted) but it is probably not superior for highly sensitive detection of HBV infectivity than the rather insensitive primary hepatocyte cultures. Furthermore, handling of this system seems to be difficult.

A most recent development uses HBV core particles derived from stably transfected Hep G.2.215 cells or from HBV carrier plasma. These particles can be introduced into permissive, but non-susceptible hepatoma cells, e. g. Huh7, by the aid of a lipofection system (Bioporter). This leads to a hundred times more efficient transport of the HBV genome to the nucleus than classical infection and to a truly productive “lipo-infection” with more progeny than input virus (B. Rabe and M. Kann, manuscript submitted). While this system is obviously not suitable for detecting inactivating or neutralisating agents targeted to the viral envelope, it may be the method of choice for validating inactivation or inhibition targeted to all other viral structures. However, this system has not yet been applied to such studies.

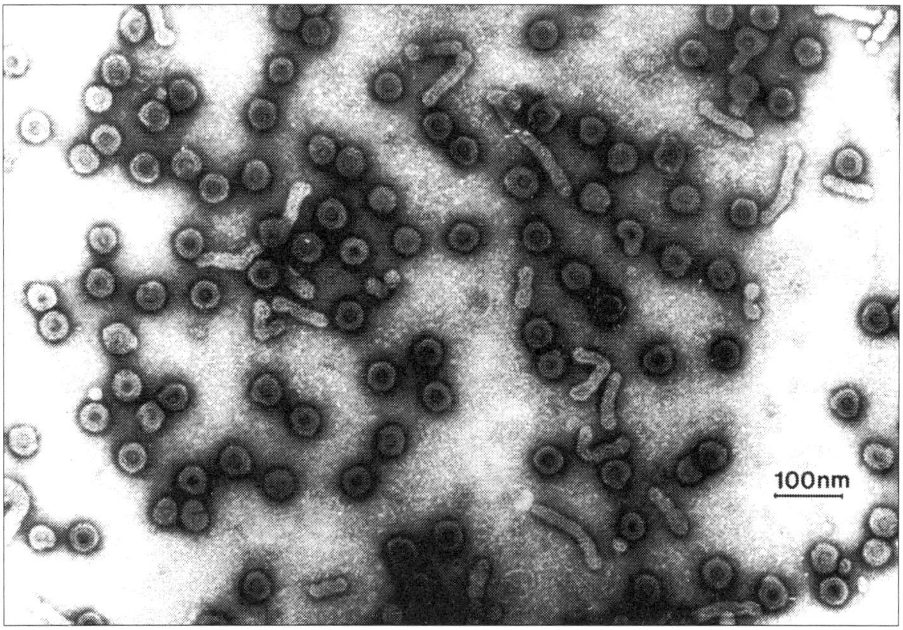

Fig. 4: Electron micrograph of HBV particles from HBV carrier plasma purified and enriched to c. 5×10^{10}/ml by gradient ultracentrifugation as described in reference 12. HBsAg particles and serum proteins are removed by a factor of > 1000.

ANIMAL HEPADNAVIRUSES AS MODELS FOR HBV

Different HBV-like virus species have been detected in woodchucks (marmots of the Eastern U.S.A.), ground squirrels, woolly monkeys (new world primates) and various old world primates (chimpanzee, orangutan, gorilla and gibbon). The latter virus isolates are considered to form one species, together with the various genotypes of human HBV [14]. There is probably no species barrier between the human and the various ape viruses. The woolly monkey virus grows, however, inefficiently in chimpanzees [15], but is highly active in Tupaia hepatocyte cultures [11]. The mammalian hepadnaviruses are grouped into the genus orthohepadnavirus. They differ no more than 40 % in the nucleotide sequence, and are very similar in genome organisation, replication and pathogenicity. Thus, their relationship to the human virus is quite close and extrapolation of results obtained with animal orthohepadnaviruses to human HBV is plausible. However, none of these animals is easy to keep in captivity or readily available. The woodchuck and its hepadnavirus have been used for studies on oncogenicity, pathogenesis, drug development and vaccination, but no reports on its use in validation of hepadnavirus inactivation have been published.

An important and more feasible animal model is the Pekin duck and its avihepadnavirus DHBV. Originally, infection studies have been done with freshly hatched ducklings, but embryonated eggs are a suitable and more convenient source for primary duck hepatocytes. However, care has to be taken that the eggs come from a DHBV-negative mother. The titres of infectivity may reach ID50 > 10^7/ml, thus allowing highly sensitive detection of residual infectivity [16]. Since all hepadnaviruses are non-cytopathic, viral gene expression or genome replication has to be monitored in the inoculated cell cultures. Detection of DHBV surface antigen in the inoculated hepatocyte cultures using a polyvalent rabbit antiserum and immunofluorescence is a feasible readout method (C. Jursch, O. Thraenhart, D. Glebe, W. Gerlich unpublished) which may be used for routine testing. An alternative is quantitative monitoring of the newly formed DHBV either with Southern blot [17] or by real-time PCR with SyBr green [18] or with specific fluorescent probes (U. Wend and W. Gerlich unpublished).

This system had been used for the validation of inactivation of DHBV in human platelets concentrates by uv-light and psoralen [19] or in red blood cell concentrates by the DNA-destroying agent PEN 110 [20]. Furthermore, various disinfectants have been validated with DHBV [16–18,21]. DHBV seems to be a more adequate model virus for HBV than other enveloped viruses. Sindbis virus and herpes virus have been used as surrogate viruses for validation of solvent-detergent virus [22] but these viruses barely resemble HBV structurally. However, it should be kept in mind that DHBV has very little sequence identity with HBV. HBV has e.g. more disulphide crosslinks between its surface proteins than DHBV and may thus be more resistant to mild detergents than HBV. Many disinfectants contain aldehydes which react preferably with lysin side chains. DHBV contains a lys-lys-lys motif on its surface whereas HBV does not. Inactivating substances targeted to viral proteins may indeed act quite differently on DHBV and HBV although no frank contradictions have yet been observed. More predictable is the effect of inactivating procedures acting on viral DNA. In this case an extrapolation of DHBV data to HBV appears justified [19,20].

Surrogate Markers for HBV Infectivity

Soon after the discovery of HBV, an attempt was made to validate inactivation of HBV by various surrogate markers. Among these were (i) the reactivity of HBsAg in commercially available immunoassays [23], (ii) the reactivity of the HBV endogenous DNA polymerase [24] and (iii) the morphological alternation and disintegration test (MADT) of the virus particles in the electron microscope [17,25]. Except for the HBsAg assay, the other markers have been compared with the infectivity in chimpanzees. However, with today's demands on the reliability of validation, these surrogate markers lack sensitivity, accuracy and the clear connection to infectivity. It is e.g. not clear whether the epitopes detected by the usual HBsAg assays are necessary for infectivity or whether infectivity is much more sensitive against inactivation than HBsAg. In a more recent approach, Jursch et al validated disinfectants in an immune assay for infectivity-neutralising epitopes in the preS1 and HBsAg domains of HBV [26]. They were able to follow the kinetics of epitope inactivation over a range of 3,000–10,000-fold inactivation and noted huge differences between purified HBsAg particles and crude HBV carrier serum. Inactivation of neutralising epitopes should theoretically be correlated to infectivity.

Another approach is the assay of genome integrity. Theoretically only virus particles containing DNA suitable for replication by DNA polymerases in vitro should be infection-competent. With today's extremely sensitive PCR techniques it is possible to follow inactivation down to very low levels of residual genomes. Of course, this approach is only useful for inactivation procedures which are targeted to the viral DNA [27]. Furthermore, the sensitivity of the assay depends on the length of the amplified DNA segment [26]. Theoretically one hit per genome is sufficient to destroy the infectivity, but usually amplified DNA products cover only a small fraction of the genome length. On the other hand, sensitivity of PCR decreases with the length of the product as shown in Figure 5 [26]. Single-stranded DNA as in the X region of the virion-associated HBV genome may be less sensitive to inactivating agents than double-stranded DNA. In Figure 5 two DNA inactivating agents are compared and validated with a PCR covering the entire HBV genome and another PCR covering the X region (U. Wend, W. Gerlich, unpublished results). Agent 1 of Figure 5 was virtually ineffective against the X region but very effective against the entire DNA genome. Such a strong difference was not seen with agent 2.

Another point of concern is that the damaged DNA genome may be repaired in the cell and become replication-competent again. This problem has been studied by the aid of DHBV for which both PCR assays and infectivity titrations are feasible. Using psoralen-induced photoinactivation [19] or a DNA-destroying chemical [20], two groups found an excellent agreement between PCR and infectivity titres for DHBV and strong activity of their agents on the PCR reactivity of HBV DNA [19,20]. Furthermore, the activity of psoralen had been previously validated in chimpanzee experiments [28]. Thus, for DNA-targeting agents, PCR seems to be a simple and sensitive method as far as inactivation of hepadnaviruses is concerned.

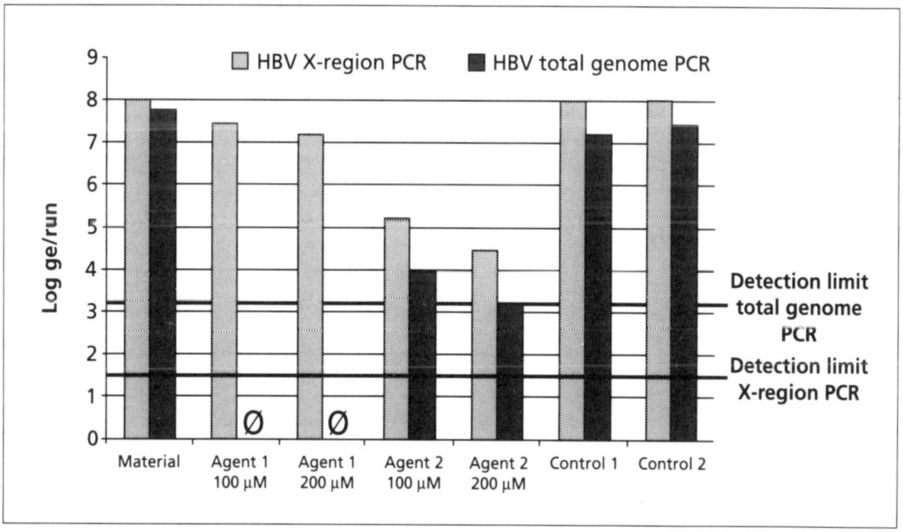

Fig. 5: Inactivation of HBV genomes in HBV carrier plasma by two nucleic acid destroying agents. HBV carrier plasma ("material") was treated according to the instructions for agents 1 or 2 or incubated in parallel without agent (controls). HBV DNA was extracted and tested as described in ref. 26.

CONCLUSIONS

The demands on the accuracy and quantitative range of a validation concerning HBV inactivation vary with the medical problem to be encountered with inactivation. Blood products will be screened for HBsAg and products from pooled donations will, furthermore, be tested for HBV DNA. Thus, heavy contaminations of the final product by HBV should be virtually impossible. The last HBV accident was caused in 1994 by a human plasma-derived product that was validated in the chimpanzee model to be free of HBV after a heat inactivation step [29,30]. In retrospect it appears that this accident was caused by a combination of various technical problems during production but not by an incorrect validation of heat inactivation. Nevertheless, even with the "gold standard" of chimpanzee titration [8], the heat inactivation of HBV has led to conflicting results of validation with very unsatisfactory results for lyophilised products [31] and incomplete inactivation in the liquid state [7]. Given the very limited number of possible inoculations and the ethical restrictions in using these animals, the future availability of these animals for HBV inactivation studies appears virtually impossible. DHBV has been accepted in Germany for validation of disinfectants against enveloped viruses [32] and has convincingly contributed to the validation of virus nucleic acid inactivation in blood products [19,20]. With other target structures of inactivation, DHBV may give a hint, but HBV should also be used in one of the cell culture models described above. Their limited capacity to detect residual infectivity may be compensated by the fact that in blood products, residual infectivity of the starting material before inactivation will be very low due to donor screening. Thus, safety of the final product may be ascertained, even in spite of large inoculated volumes. For validation of disinfectants, the HBV infectivity to be inactivated may be much higher, but the possible inocula are much smaller. Thus, proven inactivation by a reduction factor of four is considered sufficient and this can be reached with cell culture systems for HBV. Validation may be complemented by the proven destruction of neutralising epitopes. Thus, a battery of new tools for the validation of HBV inactivation is available. The use of unrelated enveloped viruses as models for HBV is no longer justified.

ACKNOWLEGEMENT

We thank M. Kann and S. Urban for communicating their submitted manuscripts before publication. The work of the authors was supported by DFG grant SFB 535/A2.

REFERENCES

1 Minegishi K, Yoshikawa A, Kishimoto S, Yugi H, Yokoya N, Sakurada M, Kiyokawa H, Nishioka K: Japanese Red Cross NAT Screening Research Group: Superiority of minipool nucleic acid amplification technology for hepatitis B virus over chemiluminescence immunoassay for hepatitis B surface antigen screening. Vox Sang 2003;84(4):287-291.

2 Meisel H, Enders A, Walther H, Wend UC, Gerlich WH: Transmission of hepatitis B virus two months prior to HBsAg positivity of donor blood. Transfusion Medicine and Hemotherapy 2003;30:228-231.

3 Roth WK, Weber M, Peterson D, Drosten C, Buhr S, Sireis W, Weichert W, Hedges D, Seifried E: NAT for HBV and anti-HBc testing increase blood safety. Transfusion 2002;42(7):869-875.

4 Kleinman SH, Kuhns MC, Todd DS, Glynn SA, McNamara A, DiMarco A, Busch MP: Retrovirus Epidemiology Donor Study: Frequency of HBV DNA detection in US blood donors testing positive for the presence of anti-HBc: implications for transfusion transmission and donor screening. Transfusion 2003;43(6):696-704.

5 Coleman PF, Chen YC, Mushahwar IK: Immunoassay detection of hepatitis B surface antigen mutants. J Med Virol 1999;59(1):19-24.

6 Westhoff TH, Jochimsen F, Schmittel A, Stöffler-Meilicke M, Schäfer JH, Zidek W, Gerlich WH, Thiel E: Fatal hepatitis B virus reactivation by an escape mutant following rituximab therapy. Blood 2003;102:1930.

7 Shikata T, Karasawa T, Abe K, Takahashi T, Mayumi M, Oda T: Incomplete inactivation of hepatitis B virus after heat treatment at 60° C for 10 hours. J Infect Dis 1978;138(2):242-244.

8 Tabor E, Purcell RH, London WT, Gerety RJ: Use of and interpretation of results using inocula of hepatitis B virus with known infectivity titers. J Infect Dis 1983;147(3):531-534.

9 Gripon P, Diot C, Guguen-Guillouzo C: Reproducible high level infection of cultured adult human hepatocytes by hepatitis B virus: effect of polyethylene glycol on adsorption and penetration. Virology 1993;192(2):534-540.

10 Galle PR, Hagelstein J, Kommerell B, Volkmann M, Schranz P, Zentgraf H: In vitro experimental infection of primary human hepatocytes with hepatitis B virus. Gastroenterology 1994;106(3):664-673.

11 Köck J, Nassal M, MacNelly S, Baumert TF, Blum HE, von Weizsäcker F: Efficient infection of primary tupaia hepatocytes with purified human and woolly monkey hepatitis B virus. J Virol 2001;75(11):5084-5089.

12 Glebe D, Aliakbari M, Krass P, Knoop E, Valerius KP, Gerlich WH: PreS1 antigen dependent infection of tupaia hepatocyte cultures with human hepatitis B Virus J Virol 2003;77(17):9511-9521.

13 Gripon P, Rumin S, Urban S, Le Seyec J, Glaise D, Cannie I, Guyomard C, Lucas J, Trepo C, Guguen-Guillouzo C: Infection of a human hepatoma cell line by hepatitis B virus. Proc Natl Acad Sci U S A 2002;99(24):15655-15660.

14 Mason WS, Burrell CJ, Casey J, Gerlich WH, Howard CR, Kann M, Newbold J, Schaefer S, Taylor JM, Will H: Family hepadnaviridae, in VIIIth Report of the International Committee for Viral Taxonomy; in press.

15 Lanford RE, Chavez D, Brasky KM, Burns RB 3rd, Rico-Hesse R: Isolation of a hepadnavirus from the woolly monkey, a New World primate. Proc Natl Acad Sci U S A 1998;95(10):5757-5761.

16 Vickery K, Deva AK, Zou J, Kumaradeva P, Bissett L, Cossart YE: Inactivation of duck hepatitis B virus by a hydrogen peroxide gas plasma sterilization system: laboratory and 'in use' testing. J Hosp Infect 1999;41(4):317-322.

17 Prince DL, Prince HN, Thraenhart O, Muchmore E, Bonder E, Pugh J: Methodological approaches to disinfection of human hepatitis B virus. J Clin Microbiol 1993;31(12):3296-3304.

18 Wang CY, Giambrone JJ, Smith BF: Comparison of cell culture systems for duck hepatitis B virus using SyBr green quantitative PCR. J Virol Methods 2002;106(2):175-184.

19 Eble BE, Corash L: Duck hepatitis B virus inactivation and 8-methoxypsoralen photoadduct formation in human platelet concentrates. Photochem Photobiol 1998, Jun;67(6):700-713.

20 Aytay S, Ohagen A, Busch MR, Alford B, Chapman JR, Lazo A: Development of a sensitive PCR inhibition method to demonstrate HBV nucleic acid inactivation. Transfusion 2004;44(4):476-484.

21 Wang CY, Giambrone JJ, Smith BF: Comparison of cell culture systems for duck hepatitis B virus using SyBr green quantitative PCR. J Virol Methods 2002;106(2):175-184.

22 Murozuka T, Aoki M, Kimura N, Sotoyama K, Abe I, Izumi H, Emura H, Takeda Y, Nanakawa H, Katsubayashi Y, Takeuchi T, Ito H, Hirakawa S, Mitsunaga S, Tadokoro K, Kanemitsu K, Miyamoto M, Yamanaka R, Nishioka K: Removal and inactivation of hepatitis B virus from contaminated pooled plasma in a large-scale manufacturing process for factor VIII and human serum albumin. Vox Sang 1999;76(3):181-186.

23 Frösner G, Jentsch G, Uthemann H: Destroying of antigenicity and influencing the immunochemical reactivity of hepatitis B virus antigens (HBsAg, HBcAg and HBeAg) through disinfectants – a proposed method for testing (author's transl). Zentralbl Bakteriol Mikrobiol Hyg [B] 1982;176(1):1-14.

24 Howard CR, Dixon J, Young P, van Eerd P, Schellekens H: Chemical inactivation of hepatitis B virus: the effect of disinfectants on virus-associated DNA polymerase activity, morphology and infectivity. J Virol Methods 1983;7(3):135-148.

25 Thraenhart O, Kuwert EK, Scheiermann N, Dermietzel R, Paar D, Maruhn D, Alberti A, Richter HJ, Hotz J: Comparison of the morphological alteration and disintegration test (MADT) and the chimpanzee infectivity test for determination of hepatitis B virucidal activity of chemical disinfectants. Zentralbl Bakteriol Mikrobiol Hyg [B] 1982;176(5-6):472-484.

26 Jursch CA, Gerlich WH, Glebe D, Schaefer S, Marie O, Thraenhart O: Molecular approaches to validate disinfectants against human hepatitis B virus. Med Microbiol Immunol (Berl) 2002;190(4):189-197.

27 Hilfenhaus J, Gröner A, Nowak T, Weimer T: Analysis of human plasma products: polymerase chain reaction does not discriminate between live and inactivated viruses. Transfusion 1997;37(9):935-940.

28 Alter HJ, Creagan RP, Morel PA, Wiesehahn GP, Dorman BP, Corash L, Smith GC, Popper H, Eichberg JW: Photochemical decontamination of blood components containing hepatitis B and non-A, non-B virus. Lancet 1988;24-31;2(8626-8627):1446-1450.

29 Clemens R, Weidmann E, Merkle W, Bock HL, Mauler R, Hilfenhaus J: Virus safety of a pasteurized antithrombin III concentrate. Preclinical and clinical data. Arzneimittelforschung 1987;37(7):759-762.

30 Mauler R, Merkle W, Hilfenhaus J: Inactivation of HTLV-III/LAV, hepatitis B and non-A/non-B viruses by pasteurization in human plasma protein preparations. Dev Biol Stand 1987;67:337-351.

31 Purcell RH, Gerin JL, Popper H, London WT, Cicmanec J, Eichberg JW, Newman J, Hrinda ME: Hepatitis B virus, hepatitis non-A, non-B virus and hepatitis delta virus in lyophilized antihemophilic factor: relative sensitivity to heat. Hepatology 1985;5(6):1091-1099.

32 Stellungnahme des Arbeitskreises Viruzudie beim Robert Koch-Institut (RKI) sowie des Fachausschusses Virusdesinfektion der Deutschen Gesellschaft zur Bekämpfung der Viruskrankheiten (DVV) und der Desinfektionsmittelkommission der Deutschen Gesellschaft für Hygiene und Mikrobiologie (DGHM): Prüfung und Deklaration der Wirksamkeit von Desinfektionsmitteln gegen Viren. Bundesgesundheitsblatt- Gesundheitsforsch- Gesundheitschutz 2004;47:62-66.

Dr. W.H. Gerlich, Institute for Medical Virology, Universitätsklinikum Giessen, Frankfurter Strasse 107, D-35392 Giessen, Germany.
E-mail: wolfram.h.gerlich@viro.med.uni-giessen.de

Robertson J (ed): PDA/EMEA European Virus Safety Forum.
Dev Biol. Basel, Karger, 2004, vol 118, p 123.

SESSION IV

Virus Inactivation/Removal technologies: Application and Validation

Chairs: *J. Löwer* (PEI)
E.-G. Graf (Pall)

Robertson J (ed): PDA/EMEA European Virus Safety Forum.
Dev Biol. Basel, Karger, 2004, vol 118, pp 125-127.

Robustness Studies on Virus Inactivation: Definition of Parameters and their Generic Use

H. Dichtelmüller

Virus Validation, Biotest Pharma GmbH, Dreieich, Germany

INTRODUCTION

Investigation of robustness of virus inactivation/removal steps during manufacture of plasma derivatives is regulated in CPMP/BWP guidelines 268 and 269. Reference to robustness is more or less randomly spread over the text, but no real guidance is available.

However, robustness studies are the most work-intensive part of all virus validation studies and need more attention than currently given. The following presentation indicates problems with robustness studies and proposes solutions for an improved approach to robustness studies without risking the viral safety of the product.

DEFINITION OF ROBUSTNESS

Robustness parameters are defined by the limits of the production process, i.e. upper and lower limit of parameters such as temperature. Robustness is not defined by the theoretical limits of the process itself and robustness studies are not performed to discover process limits which have nothing to do with the production process.

Process parameters can also be defined by analytical methods used for in-process controls of production parameters (e.g. protein content or TNBP/Tween concentration).

CRITICAL PARAMETERS

Critical parameters are process-specific. In general, critical parameters are the concentration of the virus inactivating agent, pH, temperature and protein content.

Robustness studies therefore should consider the upper and lower limit of those parameters. More critical parameters must be considered for virus removal processes such as nanofiltration (ratio of volume/filtrate, flow rates, pressure) or, even worse, column chromatography (volume per gel, age of gel, flow rates, column dimensions, impurities, cutting of peaks and sanitization). Column chromatography therefore needs extreme efforts for validation, which should be avoided whenever possible.

Choice of Parameters and Viruses

In general, for most virus inactivation processes worst-case parameters can be assumed, i.e. low temperature, neutral pH, low concentration of inactivating agent. Limits of parameters, known or assumed as best case, should not be investigated; therefore a combination of assumed worst-case conditions can be tested for preference. This helps considerably to reduce the number of validation studies on robustness.

Another critical parameter is the choice of viruses, which should take the most resistant one or the smallest one for nanofiltration into consideration; e.g. Pseudorabies virus for TNBP/Tween or Parvovirus for nanofiltration.

Extent of Studies

Robustness studies investigating the upper and lower limits of four parameters with two test viruses have resulted in a total of 18 studies. Confirmation studies on robustness multiply the workload. For example, duplicate studies on robustness for dry heat treatment of FVIII then require 36 studies, consume FVIII activity equal to > 1300 litres of plasma and take > 58 weeks of laboratory work, blocking all other work. The workload of robustness studies consequently must be reduced to a tolerable extent.

A reduction in the number of robustness studies can be achieved by focusing on parameters, representing assumed worst-case conditions, or on a combination of worst-case conditions. Testing best-case conditions makes no sense.

Studies should be performed with the most resistant virus only, as additional studies using less resistant viruses give no additional information. Duplicate studies should be avoided; only in the case of unexpected results or significant deviation from the nominal virus reduction are additional studies required.

Robust Virus Inactivation Procedures

Well known robust virus inactivation procedures, i.e. TNBP/Tween treatment are in use worldwide by a large number of companies for a variety of products. However, the standard procedure is more or less identical for different companies and for different products. Robustness studies are uniformly and mechanically performed by every manufacturer for similar products, under similar conditions, using the same spectrum of test viruses and producing nearly identical results of virus reduction.

It should therefore be considered whether those worldwide robustness studies which have been repeated again and again for such a well known and approved virus inactivation procedure as TNBP/Tween treatment are really necessary, or whether alternatively the efforts could be used for more urgent problems such as the investigation of new virus inactivating procedures, inactivation of non-enveloped viruses, emerging viruses or sanitization of equipment etc.

Another candidate for the reduction of robustness studies were the pasteurization of human albumin, where the process is exactly and thoroughly defined to the last detail by national and international pharmacopoeias.

Conclusions

As mentioned, there are well known robust and effective virus inactivation procedures in use worldwide (e.g. TNBP/Tween treatment, pasteurization of human albumin). The extent of robustness studies for those procedures can and should be reduced, in the case where standard procedures are used. The extent of robustness studies can be reduced by testing for worst-case conditions and combined worst-case conditions, using the most robust virus and by avoiding duplicate or triplicate studies.

It is extremely important to plan robustness studies carefully and to discuss the intended studies with the regulatory authorities before starting experimental work.

Dr. H. Dichtelmüller, Virus Validation, Biotest Pharma GmbH, Landsteinerstrasse 5, D-63303 Dreieich, Germany.
E-mail: herbert_dichtelmueller@biotest.de

Robertson J (ed): PDA/EMEA European Virus Safety Forum.
Dev Biol. Basel, Karger, 2004, vol 118, pp 129-131.

PDA Virus Filter Task Force Update

G. Sofer

BioReliance, Rockville, MD, U.S.A.

Key words: Virus filters, standard nomenclature

Abstract: The PDA Virus Filter Task Force has made significant progress in producing a method that will standardize the nomenclature for virus-removal filters that remove large viruses. A method for preparation of the model bacteriophage, PR 772, has also been produced and will shortly undergo a feasibility study. A technical report on the use of virus removal filters is also being prepared.

INTRODUCTION

The use of virus removal filters has enhanced the safety of biotechnological/ biological products. Multiple vendors produce virus removal filters, but there has been no standardized nomenclature system for these filters. The need for consistency in naming this type of filter was expressed by industry at several conferences and brought to the attention of the PDA's Scientific Advisory Board. A task force, Task Force for Virus Filter Challenge and Nomenclature Standardization, was formed to address this need. The scope statement read as follows:

The goal of this task force is to develop a common nomenclature and a standardized test method to be used by filter manufacturers for classifying and identifying viral-retentive filters. Currently, the nomenclature for filters used for virus removal is inconsistent and not informative for industry. Additionally, various test methods are used to evaluate the viral retention capabilities of filters. Bacteriophages are commonly used by filter manufacturers to evaluate size-exclusion properties of their filters. Agreement will be needed on which phage and how to perform the tests. Each pharmaceutical manufacturer will still have to assess the capability of the filters for their specific application.

DISCUSSION

The task force comprises members representing the filter vendors, biopharmaceutical and pharmaceutical manufacturers, regulatory agencies, and contract testing laboratories. Not surprisingly, there were initially many opinions on how to design an appropriate method and some disagreement on the feasibility of the proposed activities. It was imperative that the Task Force made very clear to all that the method would in no way be a substitute for the evaluation or validation of virus removal filter performance under actual user conditions and that the intention was not to compare filters from one manufacturer to another.

It was recognized that, initially, it would be unrealistic to address a standardized method and nomenclature system for small viruses, since there is more variability and the larger virus removal filters were selected. It was further agreed that the purpose of the method is not to test filters at maximum or worst case operating conditions.

The filter vendors have extensive knowledge of their filter capabilities. It was agreed that they would pick parameter set points within reasonable limits for the individual filters. Ranges would be established that are based on industry practice, published literature, and the filter manufacturers' recommendations. Bracketed ranges would be selected to allow testing at conditions recommended by filter manufacturers that are realistic for commercial operations. Brackets would be defined for total throughput volume, trans-membrane pressure in a constant pressure mode or flux in a constant flow mode, and wash volume. The filter companies could then select parameter set points within reasonable limits for specific protocols from the bracketed ranges. Furthermore, the filter companies would propose scaled-down models for the performance of the method, and they would test no fewer than three different membrane lots that are released by QC and are not from the same roll of membrane.

The agreements discussed above were not reached easily. As in any consensus task force with highly knowledgeable participants, there are differences of opinion. A consensus was reached by issuing a memorandum of understanding (MOI) ['I' equals 'Intent'?]. The MOI was written by Kurt Brorson from the U.S. FDA, and it proved to be a good model for future task forces challenged with controversial issues.

To use a standardized method for naming the virus removal filters, it is imperative that the bacteriophage preparation is defined. The bacteriophage PR772 was selected by the task force as the model virus. PR772 can be purchased from ATCC or from the Felix d'Herelle Reference Center for Bacterial Viruses, Laval University, Quebec, Canada. Preparation options include sterile filter alone or sterile filter with further purification and concentration with caesium chloride density gradient ultracentrifugation. At present, the caesium chloride preparation has been evaluated by light scattering and found to be mono-disperse; a study will be carried out to evaluate the mono-dispersity of the preparation using only sterile filtration.

Conclusions

The bacteriophage preparation and the nomenclature test methods have been written and approved by the Task Force. Major contributors to this effort are listed in Table 1. Funding is anticipated for early 2004 to enable a method feasibility study before publishing the method and its subsequent implementation by the filter companies. The use of this method should not only provide consistency in the ratings of virus removal filters but may prove to be useful in trouble-shooting out of specification results.

The Task Force is moving ahead with the production of a Technical Report (TR) fashioned after PDA's TR 26: Sterilizing Filtration of Liquids. Future activities will, hopefully, include a method for nomenclature of the virus removal filters that exclude smaller viruses.

Working in a consensus group is a valuable exercise. There is an opportunity to learn from one's colleagues and to provide a contribution that can benefit others. The concept of using a memorandum of understanding was shown to be a valuable tool for reaching a consensus.

Table 1: Task Force method subcommittee members.

Kurt Brorson, U.S. FDA
Hazel Aranha, Pall Corporation
Jeff Carter, Millipore Corporation
Marina Korneyeva, Bayer Corporation
Sakae Satoh, Asahi
Sri Sundaram, Schering Plough
Hannelore Wilkommen, Clearant
Peter Wojciechowski, Centocor/J&J

Dr. G. Sofer, BioReliance, 14920 Broschart Road, Rockville, MD 20850, U.S.A.
E-mail: gsofer@bioreliance.com

Robertson J (ed): PDA/EMEA European Virus Safety Forum.
Dev Biol. Basel, Karger, 2004, vol 118, pp 133-138.

Gamma Irradiation of Intravenous Immunoglobulin

W.N. Drohan, S.I. Miekka, Y.V. Griko, R.-Y. Forng, R.E. Stafford, C.R. Hill, D.M. Mann, W.H. Burgess

Clearant, Inc., Gaithersburg, MD, U.S.A.

Keywords: Gamma radiation, virus inactivation, human, plasma, immunoglobulin, intravenous.

Abstract: This paper describes gamma irradiation of a biotherapeutic product under conditions (the Clearant Process") that protect proteins and foster inactivation of viruses and other pathogens. The treated product was immunoglobulin paste from cold ethanol fractionation of human plasma, a process intermediate in the production of intravenous immunoglobulin (IGIV). The frozen paste was irradiated on dry ice to 45 kGy, conditions that inactivate $\geq 4 \log_{10}$ of non-enveloped viruses and $\geq 6 \log_{10}$ of enveloped viruses. When IGIV purified from the irradiated paste was characterized, no protein aggregation, fragmentation, oxidation or denaturation was detected and Fab functionality remained intact.

Introduction

Gamma radiation inactivates all types of pathogens but, when applied in an uncontrolled way, may also damage proteins [1]. The Clearant Process® delivers high doses of gamma radiation under tightly controlled conditions of rate, dose, temperature and stabilizers, when necessary [2-4]. These conditions minimize protein damage, while maintaining efficient inactivation of pathogens. The Clearant Process® can be applied to final products, or can be used to inactivate pathogens in process intermediates, as described here for frozen immunoglobulin paste.

Materials and Methods

Proteins

Immunoglobulin (Ig) paste was a product of cold ethanol fractionation of pooled human plasma. IGIV was purified from the paste by conventional anion exchange chromatography [4].

Gamma irradiation

Irradiation was carried out on an immunoglobulin paste fraction obtained from cold ethanol fractionation of pooled human plasma. The paste was stored frozen at -80ºC and was irradiated on dry ice in a commercial Cobalt-60 irradiator or in an MDS Nordion Gammacell 220 research irradiator to total radiation doses of 45 or 50 kGy.

Electrophoresis

The crude Ig paste, both unirradiated and irradiated, and the IGIV purified from the paste were analysed by SDS-PAGE using the method of Laemmli [5] under non-reduced conditions in a 4-20% gradient gel and under reducing conditions in a 12% gel.

Size exclusion HPLC

IGIV (50 µg) purified from unirradiated and irradiated Ig paste was applied to a 7.8 x 300mm Biosep S-3000 column in a Waters 2690 HPLC system and 2487 UV/visible detector. Flow rate of phosphate buffered saline, pH 6.7 (PBS), was 1 ml/min.

Nephelometry

Turbidity was measured with a DRT 100B Turbidimeter (HS Scientific) at 25ºC. The instrument was modified to hold micro (0.5ml) cuvettes, which were calibrated against 25-30ml cuvettes using standard Farmazin solution. Measurements of insoluble aggregates in IGIV solution (100mg/ml, pH 6.2) were taken every one to four days during 4ºC storage and expressed in nephelopmetric units (NTU).

Rubella-binding ELISA

Purified IGIV was tested for antigen binding activity using Rubella IgG diagnostic kits obtained from Sigma following the manufacturer's protocol.

Amino acid analysis

Samples and amino acid standards were alkylated with 4-vinylpyridine, hydrolysed in 6N HCl for 20 hours at 115ºC, and derivatized with phenylisothiocyanate. The resulting PTH amino acids were resolved and quantified using Waters PicoTag amino acid analysis system with absorbance detection at 254nm.

Viral inactivation

Porcine parvovirus (PPV), encephalomyocarditis virus (EMC) and vaccinia virus (VV) were purchased, expanded, and assayed by a $TCID_{50}$ method as previously described [6].

Results

Irradiated and non-irradiated immunoglobulin paste and IGIV purified from the paste were analysed by several methods to determine the extent to which the protein was modified by irradiation of the paste.

Figure 1 shows the results of SDS-PAGE analysis of the irradiated and non-irradiated starting material and the purified IGIV products. Little difference is detectable before and after irradiation in the starting paste and IGIV. For IGIV, under non-reducing conditions a single band near 160kDa is obtained, and two bands at approximately 55 and 30 kDa are seen after reduction.

Size exclusion chromatography of the IGIV purified from irradiated and non-irradiated paste is shown under non-denaturing conditions in Figure 2. The profile for the irradiated sample was the same as that of the non-irradiated control, indicating that neither detectable aggregation nor fragmentation occurred upon irradiation. To analyse stability and propensity to aggregate further, the IGIV was stored at 4ºC for three weeks as a 10% protein solution. Nephelometric analysis (Fig. 3) showed no radiation-induced or time-dependent change in aggregation upon storage up to 21 days.

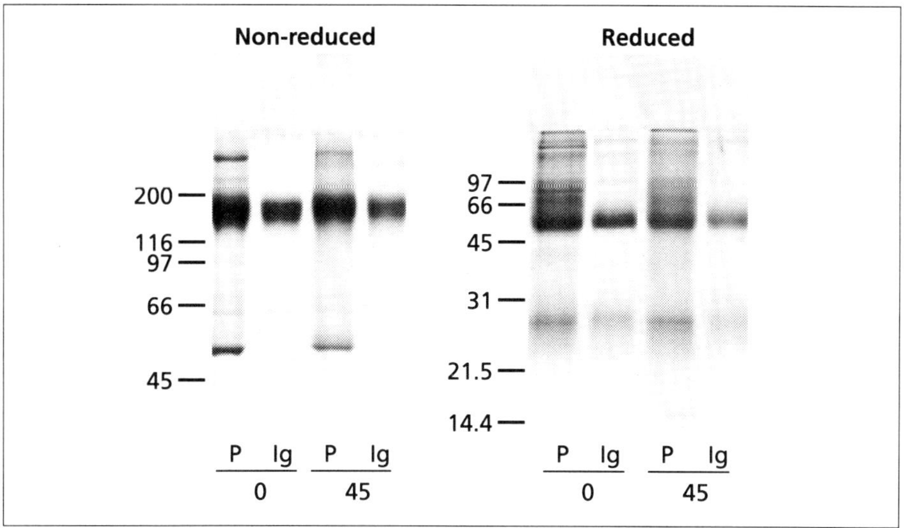

Fig 1: SDS-PAGE of non-irradiated and 45-kGy irradiated immunoglobulin paste (P) and IgG purified from the paste (IGIV).

Binding of non-irradiated and irradiated IGIV to immobilized rubella antigen in an ELISA format is shown in Figure 4. The slight shift in antigen binding activity, seen as a right-shift in the curve for the 45-kGy irradiated sample, represents less that 15% change in efficacy and is within the variability of the assay measurements. Subsequent experiments have shown 100% recovery of rubella binding after 50kGy irradiation [4].

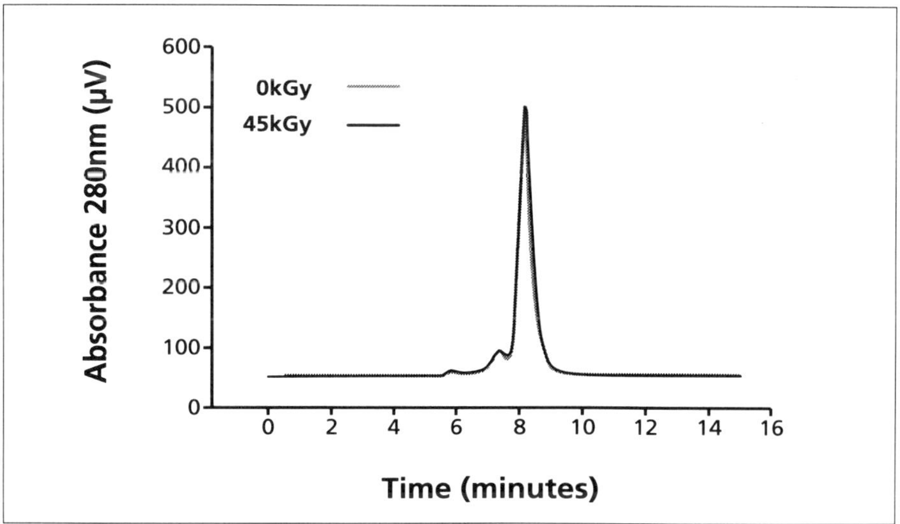

Fig. 2: Size exclusion chromatography of IGIV purified from non-irradiated and irradiated paste.

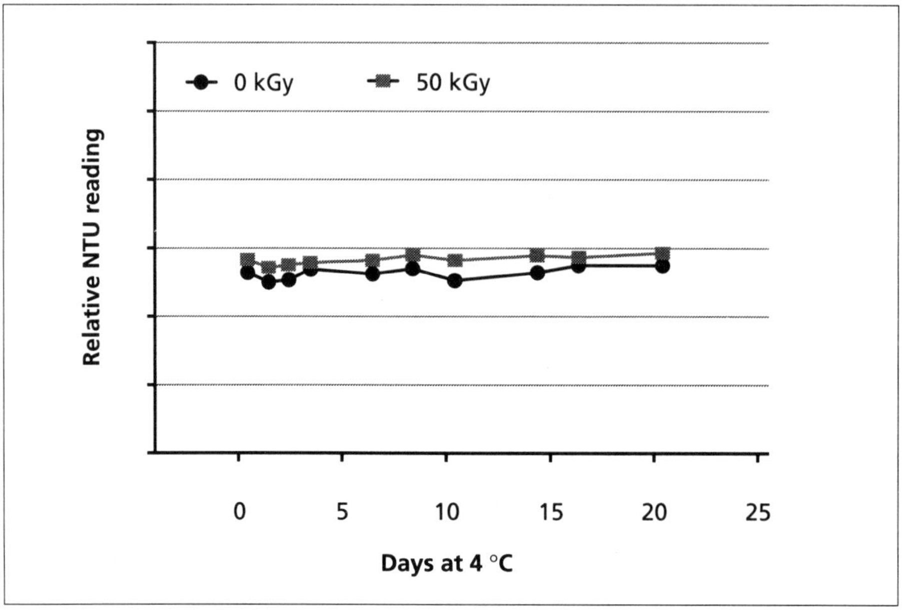

Fig. 3: Nephelometric analysis of irradiated (50-kGy) and non-irradiated IGIV during 20 days of storage at 4°C.

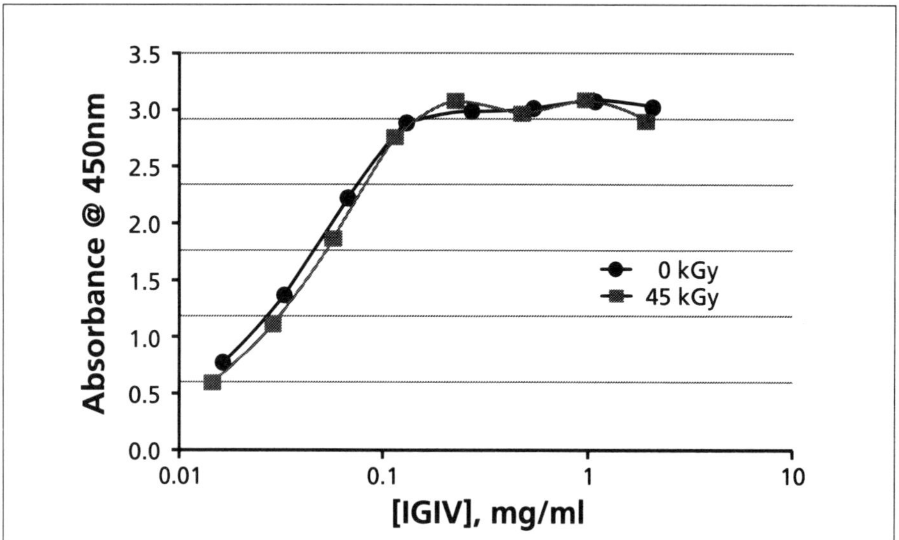

Fig. 4: Binding to rubella antigen of IGIV purified from non-irradiated and 45-kGy irradiated paste.

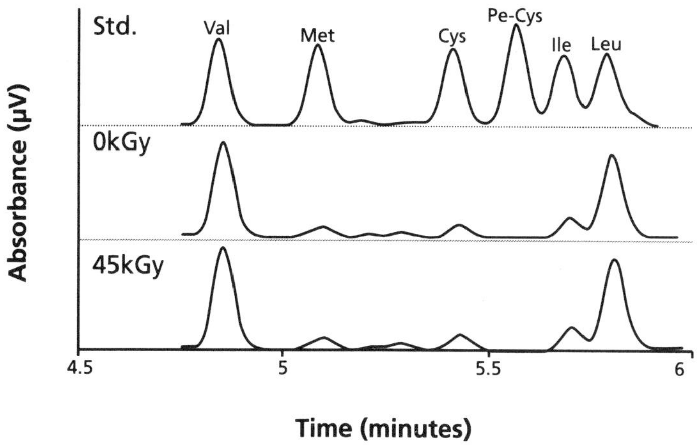

Fig. 5: Amino acid analysis of IGIV purified from irradiated and non-irradiated paste, alkylated with 4-vinyl pyridine.

The sulphur-containing amino acids are known to be prime targets for oxidative damage during gamma irradiation under uncontrolled conditions [7]. We used amino acid analysis to look for changes in these sensitive amino acids (Fig. 5). Irradiated and non-irradiated IGIV was alkylated with 4-vinyl-pyridine to convert cysteine to pyridethyl cysteine (Pe-Cys). The lack of detectable Pe-Cys in the alkylated protein indicated that no change in sulfhydryl groups occurred upon irradiation. In addition, no changes were detected in methionine or cystine content, indicating that oxidation of these susceptible amino acids did not occur upon irradiation of IGIV by the Clearant Process®.

Spectroscopic techniques were applied to assess conformational and structural changes. The near UV and far UV CD spectra were both identical for unirradiated and 45-kGy irradiated product, showing that the secondary and tertiary structures were not detectably changed (data not shown). Similarly, the first derivative UV spectra of unirradiated and irradiated IGIV were identical, indicating that the environment of the aromatic amino acid residues were unaffected by gamma irradiation (data not shown).

Virus inactivation was demonstrated in IGIV solution (Table 1). At 45 kGy without additives, the two non-enveloped viruses PPV and EMC were inactivated by 4.8 $\log_{10}$ and ≥5.8 $\log_{10}$ respectively, and the enveloped virus VV was inactivated by ≥ 5.5 $\log_{10}$.

Table 1: Inactivation of viruses in IGIV by 45 kGy of gamma radiation.

Virus	Lipid-envelope	Log_{10} reduction factor
PPV	No	4.8
EMC	No	≥5.8
VV	Yes	≥5.5

Discussion

The current studies show that gamma irradiation, under controlled conditions of dose rate and total dose with the temperature maintained near -70ºC with dry ice, allows excellent recovery of IGIV structure and function. A recent publication from our laboratory [4] further reported that in IGIV irradiated to 50 kGy under these conditions, the Fc domain was fully functional and unchanged in its ability to bind specifically to Fcg receptors and to induce cellular responses. In addition, the Fab domain maintained its ability to bind three viral antigens–rubella, CMV and mumps [4].

The Clearant Process® of pathogen inactivation has broad applicability, as illustrated by the ability to irradiate manufacturing intermediates such as pastes/precipitates from ethanol fractionation. Downstream purification steps provide the possibility to purify fully functional product, or to remove radioprotectants that may be used to optimize product recovery.

Gamma Irradiation has been shown to be very effective in inactivating both lipid enveloped and nonenveloped viruses. In addition to the inactivation of viruses in plasma pastes and purified IGIV [4], we have demonstrated that PPV and other non-enveloped viruses are inactivated in preparations of purified monoclonal antibodies [2] and urokinase [3], while protein structure and function are maintained. Consequently, gamma irradiation appears to be a robust technology that allows for the inactivation of viruses in protein-containing biologicals while providing for the high level recovery of biological activity.

References

1 Kempner ES: Effects of high-energy electrons and gamma rays directly on protein molecules. J Pharm Sci 2001;90(10)1637-1646.

2 Grieb T, Forng RY, Brown R, Owalabi T, Maddox E, McBain A, Drohan WN, Mann DM, Burgess WH: Effective use of gamma irradiation for pathogen inactivation of monoclonal antibody preparations. Biologicals 2002;30(3):207-216.

3 Amareld RW, Wersocki M, Drohan, M, Drohan WN, Burgess WH, Mann DM, Forng RY: Controlled gamma-irradiation mediated pathogen inactivation of human urokinase preparations with significant recovery of enzymatic activity. Biologicals 2003;31:261-264.

4 Tran H, Marlowe K, McKenney K, Petrosian G, Griko Y, Burgess WH, Drohan WN, Imboden MA, Kempf C, Boschetti N, Mann DM: Functional integrity of intravenous immunoglobulin following irradiation with a virucidal dose of gamma irradiation. Biologicals 2004, in press.

5 Laemmli UK: Cleavage of structural proteins during the assembly of the head of bacteriophage T4. Nature 1970;259:680-685.

6 Miekka SI, Forng RY, Rohwer RG, MacAuley C, Stafford RE, Flack SL, MacPhee M, Kent RS, Drohan WN: Inactivation of viral and prion pathogens by gamma irradiation under conditions that maintain the integrity of human albumin. Vox Sang 2003;84(1):402-408.

7 Kumta US, Shimazu F, Tappel AL: Decrease of radiation damage to proteins by sulfhydryl protectors. Radiation research 1962;16:679-685.

Dr. W.N. Drohan, Clearant, Inc., 401 Professional Drive, Gaithersburg, MD 20879, USA.
E-mail: bdrohan@clearant.com

Robertson J (ed): PDA/EMEA European Virus Safety Forum.
Dev Biol. Basel, Karger, 2004, vol 118, pp 139-147.

UV-C Irradiation of Human Albumin Solution on Manufacturing Scale

A.J. MacLeod

Protein Fractionation Centre, Scottish National Blood Transfusion Service, Edinburgh, U.K..

INTRODUCTION

Consideration of the application of UV-C irradiation to the inactivation of viruses in blood products such as human albumin solution immediately raises two questions. The first is "Why is UV-C irradiation not already in widespread use?" Its effectiveness in the inactivation of viruses has been appreciated since the 1940s [1] and the results of investigations into its use have continued to be reported up to the present [2,3] which overall represents more than 50 years of active research. This has to be contrasted with the speed with which other technologies directed at the same objective, such as solvent/detergent treatment and heat treatment, have been developed and introduced into routine use. It is the purpose of this presentation to try to address this question and to report on progress that has been made towards development of a cGMP- compliant system that could be introduced into current manufacturing processes.

The second question raised by the topic of this presentation is "Why study the treatment of human albumin solution?" Infectious contamination is not widely considered to be a major problem with the human albumin solution products that are currently available and this would not figure in most lists of products requiring urgent attention in this respect. However, human albumin does only have a single virus inactivation step in its manufacturing process, pasteurisation at +60°C for 10 hours, and there is a general move within the blood products industry towards the introduction of two complementary virus inactivation or removal steps into the manufacture of each product. On a more pragmatic level, albumin has recently been readily available to support validation studies on the application of UV-C treatment up to full manufacturing scale. It was on this basis that the SNBTS in collaboration with Iatros Ltd, (Prospect 2, Dundee Technology Park, Dundee DD2 1TY, U.K.) has undertaken a programme of work with the objective of developing a system for UV-C irradiation of human albumin solution able to:

(a) inactivate at least four logs and preferably greater than five logs of canine parvovirus (CPV) and of hepatitis A virus (HAV);

(b) treat a 1000L batch of 4.5%(w/v) human albumin solution in about four hours, i.e. a through-put of 4 to 4.5L/min, producing a clinically acceptably product;

(c) process 20%(w/v) human albumin solution and other liquid plasma protein products

Virus Inactivation

The potential for the application of UV-C treatment can be appreciated from comparison of virus inactivation data obtained for pasteurisation of human albumin solution and for UV-C irradiation. Pasteurisation has been found to be very effective against enveloped viruses and some non-enveloped viruses but some non-enveloped viruses, notably CPV and HAV, have been found to be relatively resistant to this treatment (Table 1). On the other hand both CPV and HAV have been found to be very sensitive to UV-C irradiation although it should be noticed that not all viruses are equally susceptible, as shown from our studies on the results for pseudorabies (PRV) and bovine viral diarrhoea virus (BVDV) (Table 2).

From these data it can be appreciated that UV-C is particularly effective against some small non-enveloped viruses which are relatively resistant to other inactivation treatments, such as pasteurisation, and is therefore potentially a useful procedure to complement these other treatments.

UV-C Irradiation Conditions

The UV-C light used in irradiators is produced by conventional low-pressure mercury vapour lamps and the predominant wave-length is 254nm. The challenge in using this light to inactivate viruses in protein solutions arises from the absorbance (A) of the light by the proteins themselves (Table 3).

Table 1: Inactivation of viruses during pasteurisation at +60°C.

(a) Enveloped viruses			(b) Non-enveloped viruses		
Virus	**Time point (h)***	***R***	**Virus**	**Time point (h)***	***R***
Sindbis	0	4.1	SV40	0	3.6
	2	≥5.8		2	≥4.8
SLFV	0	4.8	Polio	0	5.2
	2	≥5.8		4	≥5.4
HSV	0	4.1	CPV	0	0.0
	2	≥4.6		10	2.6
Vaccinia	0	2.3	HAV	0	0.1
	2	≥4.5		10	3.1

* T = 0 is the point when +60°C was achieved.
R = $\log_{10}$ virus reduction factor.

Table 2: Inactivation of viruses by irradiation with the same dose of UV-C.

Virus	*R*
HAV	≥5.2
	≥5.2
CPV	≥7.0
	≥7.0
PRV	2.8±0.4
	2.7±0.5
BVDV	5.3±0.5
	5.6±0.5

Table 3: Absorbance at 254nm (A_{254}) of protein solutions.

Protein solution	**Total protein**	A_{254}
(%w/v)	**(g/l)**	**(1cm)**
4.5% HA	45	11.9
20% HA	200	51.8
IV-IgG	50	27.4

HA = human albumin solution; IV-IgG = intravenous immunoglobulin solution.

The effect of absorbance is described by the Beer-Lambert law which states that light intensity in a sample decreases exponentially with sample thickness (depth) and concentration. From this the relationship can be derived:

$$A = -\log_{10} \frac{\text{(intensity of transmitted light)}}{\text{(intensity of incident light)}}$$

The consequences of this effect on the ability of UV-C irradiation to inactivate viruses in protein solution is illustrated by a study of the inactivation of phage ϕX174 in human albumin solution prepared at a range of protein concentrations (Fig. 1).

These results demonstrate that although in dilute solutions UV-C is very effective at inactivating the phage with *R* about seven logs when A_{254} is 1, when A_{254} is increased to 12, equivalent to 4.5% albumin solution, *R* is reduced to about one log. The magnitude of the observed effect in this study is in spite of the light only being required to penetrate 3 mm into the solution to irradiate the entire volume, the tube being exposed to lamps on all sides, and demonstrates the limited effect that could be expected from surface irradiation of bulk solutions. This has been appreciated since the first attempts to develop UV-C irradiated plasma and plasma protein products and consequently the emphasis has been on the design of continuous-flow devices.

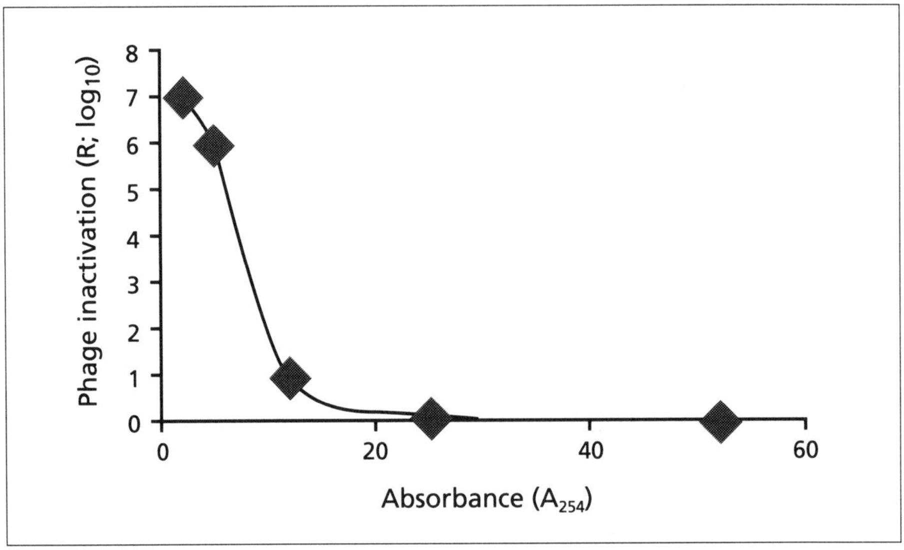

Fig. 1: Inactivation of phage ϕX174 in human albumin solution by UV-C irradiation. Samples were irradiated by passage through a silica glass tube (id = 6 mm) with four parallel UV-C lamps arranged around it. Sample residence time in the irradiator = 20 sec.

UV-C Irradiation Device Designs

Historically devices for UV-C irradiation of plasma protein solutions have concentrated on producing a thin layer of liquid of the order of 100 μm deep flowing across a surface where it was exposed directly to UV-C light [4-7]. Such devices have the advantage that they offer unimpeded exposure of the product to the light but the devices are mechanically complex, have a restricted flow-rate with a maximum of about 500 mL/min and it has been reported that the product heats up significantly during treatment [8].

Recently devices have been described with a more robust design in which the product is irradiated while it is flowing through a transparent tube. The most straight-forward of these devices relies on the product having a sufficiently low protein concentration so that it is transparent to the UV-C light which is therefore able to penetrate the entire thickness of the irradiation tube [9]. These devices have the advantage that the depth that the light is required to penetrate is not affected by the product flow-rate so that through-puts of litres per minute are quite feasible. Moreover, because of the transparency of the product, it is possible to monitor the treatment by positioning the lamp on one side of the irradiation tube with a UV-C detector diametrically opposite. However the limitation that arises with these devices is the requirement for low A_{254}. It has already been demonstrated that for this approach to be effective, an A_{254} of one or less is required. Consequently with 4.5% albumin solution, with an A_{254} of 12, a 12-fold dilution of the product would be required; thus for a 1000L batch the addition of 11,000L of water to the product would be required before the treatment which would have to be removed afterwards. Similarly, for a 250L batch of IV-IgG, the addition and subsequent removal of over 6000L of water would be required and for a 250L batch of 20% albumin it would be over 12,000L of water. Nonetheless it must be appreciated that devices using this design are readily cGMP compliant and could find application in the treatment of products that have a low protein concentration, such as plasma-derived coagulation factor preparations or recombinant protein products, or else small batch sizes where dilution may not be so much of a problem.

An alternative basis for the design of UV-C irradiators for treatment of higher protein content products is to force the entire volume of the product to pass through the very thin layer immediately inside the wall of a transparent tube where the light is sufficiently intense to be virucidal. This can be done by installing mixing elements inside the tube [10] or else by configuring the tube so that turbulence and hence mixing is induced in the fluid as it passes through [11]. These devices have the advantage that they can be used for treatment of undiluted plasma protein products such as the albumin solutions or IV-IgG in large batches at flow-rates in our experience of up to 5L/min. Their major limitation is that the product is, as already discussed, essentially opaque to the UV-C light so it is not possible to monitor the treatment by straightforward detection of transmitted light.

UV-C Irradiation Devices Using In-line Mixing: Operating Factors

Devices relying on mixing to expose an entire batch of product to virucidal UV-C intensity are also more difficult to scale-up or -down for validation purposes. For a device with mixing elements installed in the tube [10] a design equation has been developed [12].

$$R = \frac{\text{K x UV-Intensity (I) x Residence Time } (t_R) \text{ x Mixing Factor (Z)}}{\text{Fluid Absorbance } (A_{254}) \text{ x Tube Diameter (d)}}$$

The Mixing Factor (Z) is in turn a derivative of the Reynold's Number which describes the hydrodynamic behaviour of a fluid in an open tube and is calculated from the fluid density, fluid viscosity and its velocity through the tube [13]. K is an empirically determined constant which is virus-specific but which also includes factors which depend on the physical characteristics of the system. From the design equation it is possible to calculate the expected value of R for a given virus under any set of conditions and to predict the effect of varying the conditions (Fig. 2).

Similar studies have been done to validate the inactivation of mammalian viruses in a unit designed for manufacturing use and have given a calculated R for CPV and HAV in 4.5% albumin under manufacturing conditions of about 10 logs (Table 4).

These studies have demonstrated that the UV-C irradiation device with mixers inserted into the tube can be scaled-up or -down predictably and can be operated at manufacturing scale. The device in use by the SNBTS has been used to process batches of 4.5% albumin up to 700L and has been run continuously for periods of up to 12 hours.

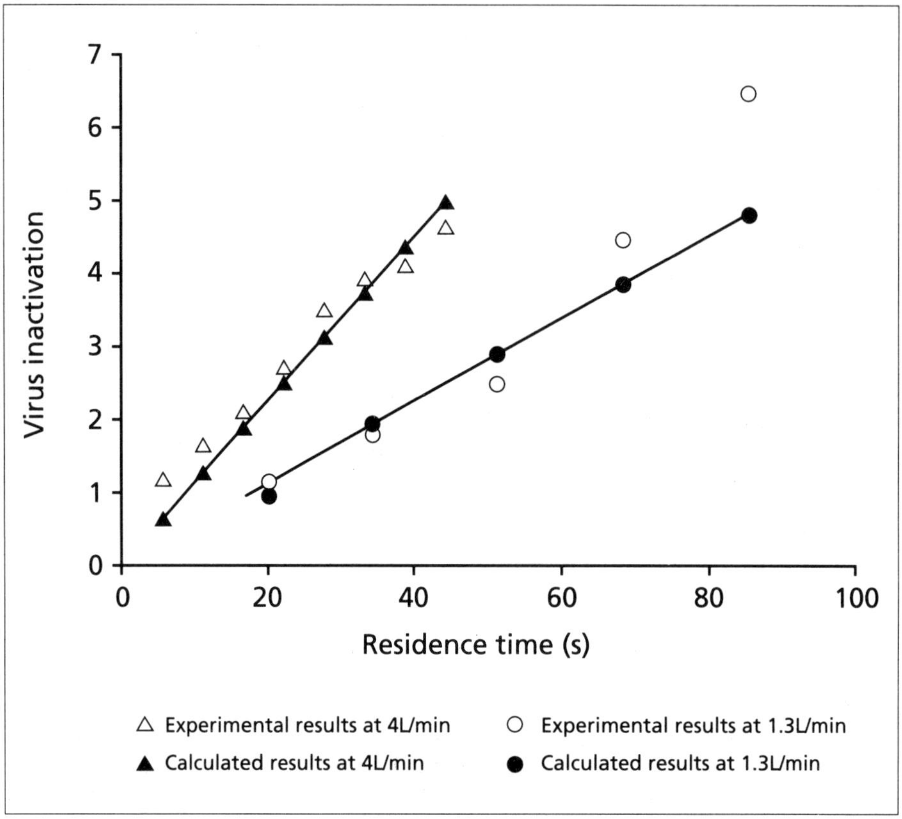

Fig. 2: Effect of variation in flow-rate and residence time on inactivation of phage ϕX174.

Table 4: Inactivation of mammalian viruses in 4.5% human albumin solution at pilot and production scales.

Irradiation zone length (m)	**Degree of virus inactivation (R; $\log_{10}$)**			
	Canine Parvovirus		Hepatitis A Virus	
	Calculated	Measured	Calculated	Measured
2.1	4.7	4.3±0.4	4.8	4.7±0.4
	(Flow-rate 5.5L/min)		(Flow-rate 4.8L/min)	
4.2	10.3	NA	10.0	NA
	(Flow-rate 4.2L/min)		(Flow-rate 4.2L/min)	

PRODUCT QUALITY

In addition to the validation of the efficacy of UV-C irradiation devices with regard to virus inactivation, questions are also raised about the quality of the product. In particular, concern is voiced about the degree of oxidative damage which may be caused to the product by the intense irradiation and in some studies the addition of anti-oxidants and free-radical scavengers to the product have been shown to protect some proteins in plasma, notably coagulation factor VIII, from inactivation [14]. Similarly it has been reported that in albumin and IV-IgG solutions irradiated in a thin film device in the absence of a protective agent, there was an increased extent of aggregation by two- to three-fold [15]. However, several of the amino acids in albumin adsorb strongly in the region of 254nm and in a study of the inactivation of enzymes by monochromatic UV light, Setlow and Doyle [16]found a sharp peak in the action spectrum at about 250nm. This is the region of an absorbance maximum for the disulphide bond in cystine and it was concluded that the observed inactivation of the enzymes was due to denaturation following cleavage of these stabilising bonds. This is a direct effect of the UV light on the protein and is not mediated by any other factors in the solution.

In studies of product quality following UV-C irradiation in a transparent tube fitted with mixing elements we have not observed any generation of aggregates in albumin after UV-C treatment alone (Table 5). However after subsequent pasteurisation, a UV-C dose related increment in the total aggregate content of the product was observed, although at the dose levels anticipated for use in manufacture of clinical products (CVP $R \cong 10$logs) the increase was much smaller than that attributable to pasteurisation alone.

Analysis of these samples by SDS-PAGE in non-reducing conditions has shown that the pattern of aggregates is very similar in all pasteurised samples with or without UV-C irradiation and that there is none detectable in reducing conditions. This is interpreted as indicating that the cause of the aggregation in these samples is almost entirely due to inter-molecular disulphide bond formation.

It has been shown that the removal of trace level glycoprotein contaminants in albumin reduces the formation of aggregates during pasteurisation [17]. Investigation of this approach to controlling aggregation shows that it is also effective in reducing aggregate formation after UV-C treatment and pasteurisation (Table 6).

Table 5: Product quality in 4.5% human albumin solution after UV-C irradiation and pasteurisation.

UV-C treatment	Monomer (% AUP)		Dimer (% AUP)		Aggregate (% AUP)	
	UV C only	UV-C + pasteurisation	UV-C only	UV-C + pasteurisation	UV-C only	UV-C + pasteurisation
None	90.5*	87.6*	6.9*	5.9*	2.6*	6.5*
x1	90.2	87.1	7.2	6.2	2.6	6.8
x2	90.1	86.1	7.3	6.4	2.6	7.4
x6	89.7	76.3	7.8	7.5	2.5	16.1

UV-C Treatment: x1 ≡ CPV $R \cong 10$logs; x2 $R \cong 20$logs; x6 $R \cong 60$logs.
%AUP = % total area under the peaks by HPLC size exclusion chromatography.
* = non-UV-C treated, pre- or post-pasteurisation sample.

Table 6: Product quality of 4.5% albumin with reduced glycoprotein content after UV-C irradiation and pasteurisation.

UV-C treatment	Monomer (% AUP)		Dimer (% AUP)		Aggregate (% AUP)	
	UV-C only	UV-C + pasteurisation	UV-C only	UV-C + pasteurisation	UV-C only	UV-C + pasteurisation
None	94.2*	93.7*	5.8*	4.7*	0.0*	1.6*
x1	94.4	93.2	5.6	5.0	0.0	1.9

* = non-UV-C treated, pre- or post-pasteurisation sample.

These results demonstrate that reduction of the glycoprotein content eliminates aggregate formation in 4.5% albumin solution after UV-C treatment alone, as well as reducing the level seen after subsequent pasteurisation. There is still some additional aggregation apparent after pasteurisation in samples that had been previously treated with UV-C but, as before, this increase is much smaller than the increase attributable to pasteurisation alone. In the case of 20% albumin it has been found that the level of aggregate formation in all samples is higher than in the corresponding samples of 4.5% albumin but that reduction of the glycoprotein content has the same effect in substantially reducing the final aggregate content.

Conclusions

From these studies it is clear that irradiation with UV-C light with a wave-length of 254nm can be an effective treatment for virus inactivation in pharmaceutical protein solutions and that the spectrum of viruses inactivated by UV-C complements that of many other treatments. Devices are now available for this process which are scaleable from 100mL to 1000L. This equipment can be constructed from readily available, robust and reliable components and can be easily integrated into existing manufacturing processes. Devices depending on low protein concentrations in the in-process material have potential use in specific applications such as treatment of coagulation factor products. However, their use for treatment of higher protein concentration products such as albumin and immunoglobulins would involve unmanageably large volumes and introduce severe water handling problems. As for the treatment of products with higher protein contents, the major processing issue remains the development of a method for monitoring the UV-C dose to achieve cGMP compliance and this is at present an area of active investigation. It has been shown that UV-C does have an effect on albumin in solution resulting in some additional aggregation, apparently primarily through absorption by cystine disulphide bonds leading eventually to intermolecular disulphide bond formation. This effect appears to be exacerbated by low-level glycoprotein contaminants in the case of albumin solution and it is substantially reduced by their removal. However at the UV-C dose range envisaged for the treatment of clinical products, the impact on product quality is much less than that caused by the current treatments which have been in use for many years.

Acknowledgements

A substantial team of people has been engaged at various times in various aspects of studies described in part in this presentation. They are too many to acknowledge them all individually but particularly notable contributions have been made by Dr Q Li, M Gray, Dr C Bienek, A Jones, M Roff, A Bartley, D Latto, Dr P Foster, Dr A Welch, Dr J Rudge and A Cavaghan (SNBTS Protein Fractionation Centre Development Department), Dr S MacDonald, Dr D Pepper, and L Morrison (SNBTS National Science Laboratory) and I Cameron and A Gunn (Iatros Ltd, Prospect 2, Dundee Technology Park, Dundee DD2 1TY, U.K.).

References

1 Wolf AM, Mason J, et al: Ultraviolet irradiation of human plasma to control homologous serum jaundice. JAMA 1947;135:476-477.

2 Hart H, Reid K, Hart W: Inactivation of viruses during ultraviolet light treatment of human intravenous immunoglobulin and albumin. Vox Sang 1993;64:82-88.

3 Chin S, Jin R, et al: Virucidal treatment of blood protein products with UVC radiation. Photochem Photobiol 1997;65:432-435.

4 Milzer A, Oppenheimer F, Levinson SO: A new method for the production of potent inactivated vaccines with ultraviolet irradiation. J Immunol 1945;50:331-340.

5 Habel K, Sockrider BT: A continuous flow method of exposing antigens to ultraviolet irradiation. J Immunol 1947;56:273-279.

6 Blanchard MC, Stokes J, et al: Ultraviolet irradiation of human plasma to control homologous serum jaundice. JAMA 1948;138:341-343.

7 Marx G, Mou X, et al: Protecting fibrinogen with Rutin during UVC irradiation for viral inactivation. Photochem Photobiol 1996;63:541-546.

8 Kallenbach NR, Cornelius PA et al: 1989; Inactivation of viruses by ultraviolet light, in Morgenthaler J-J (ed): Virus inactivation in plasma products. Curr Stud Haematol Blood Transfus. Basel, Krager, 1989, vol 56, pp 70-82.

9 Laub R, De Wael L, Di Giambattista M: US Patent no. 6,190,608 "Method and apparatus for inactivating contaminants in blood products", 2001.

10 Gunn A, Cameron ID, et al : US Patent No. 6,586,172. Device for treatment of biological fluids, 2003.

11 Kaiser K, Kauling J, et al: PCT Patent Application No. WO 02/38191. Method of inactivating microorganisms in a fluid using ultraviolet radiation, 2002.

12 Li Q, McLean C, MacDonald S, MacLeod A, Foster P: Inactivation of non-enveloped viruses by UV-C irradiation: development of a predictive model. Proc 18th Congress ISTH, Paris 2001. Abstr P2607, 2001.

13 Bird RB, Stewart WE, Lightfoot ZN: Transport phenomena. New York: John Wiley & Sons, 1960.

14 Chin S, Williams B, et al: Virucidal short wavelength ultraviolet light treatment of plasma and factor VIII concentrate: protection of proteins by antioxidants. Blood 1995;86:4331-4336.

15 Chin S, Anderson I, et al: Virucidal treatment of albumin and IVIG with UVC irradiation. 24th Cong Int Soc Blood Transf, Makuhari, Japan, 1996.

16 Setlow R, Doyle B: The action of monochromatic ultraviolet light on proteins. Biochim Biophys Acta 1957;24:27-41.

17 Jensen LB, Dam J, Teisner B: Identification and removal of polymer- and aggregate-forming proteins in human plasma albumin preparations. Vox Sang 1994;67:125-131.

Dr. A.J. McLeod, Protein Fractionation Centre, Scottish National Blood Transfusion Service, Edinburgh EH17 7QT, U.K.
E-mail: alex.macleod@snbts.csa.scot.nhs

Robertson J (ed): PDA/EMEA European Virus Safety Forum.
Dev Biol. Basel, Karger, 2004, vol 118, p 149.

SESSION V

Virus Safety Aspects of Advanced Technology Medicinal Products

Chairs: *J. Robertson* (NIBSC)
H. Willkommen (Clearant)

Robertson J (ed): PDA/EMEA European Virus Safety Forum.
Dev Biol. Basel, Karger, 2004, vol 118, pp 151-153.

Virological Issues in the Use of Cell Therapies

P.D. Minor

NIBSC, Potters Bar, Herts., U.K.

INTRODUCTION

Cell therapies for the treatment of human disease are a relatively new clinical approach with great potential for treatment of disorders which are currently unsatisfactorily managed such as Parkinson's disease or where lifetime treatment is required, as in insulin-dependent diabetes. It could be possible to introduce cells secreting specific factors or proteins to have a permanent or long-term effect rather than give repeated doses of a replacement substance. The most practically advanced applications so far are probably for skin substitutes which may have a major application in the treatment of burns, and where the desired material has a structural rather than more conventional pharmaceutical effect. The approach encompasses stem and differentiated cell therapies.

Human cells, tissues and fluids have been used in the production of biological medicines for many years, and raise specific virological issues in that all viruses which may contaminate them are ipso facto capable of infecting humans. This paper deals with the appropriateness of current approaches to virological safety with particular reference to cell therapies.

TYPES OF HUMAN MATERIAL USED IN PRODUCTION

Plasma from human sources is fractionated to give a large number of widely used medicinal products, including clotting factors, immunoglobulins and albumin. In the past this involved sourcing starting material from virologically poorly characterised sources i.e humans. Increasingly over recent years, procedures have been put in place to ensure that the plasma is free of the agents causing the most concern, such as the viruses causing hepatitis B, hepatitis C and Human Immunodeficiency Virus including testing the donors, and more recently introducing a period of quarantine of donations, only released when the donor returns for retesting. A further source of assurance of safety comes from the process used to generate the final purified product, which may be deliberately treated to remove or inactivate infectious agents in addition to any effect of the production process itself. Such treatments are unlikely to be compatible with cell viability and therefore could not be applied to cell therapies, but donor screening is clearly a major possibility.

A number of biological medicinal products are manufactured in cell culture systems involving human cells, including many viral vaccines but also other types of product, such as urokinase. It has been proposed to use human monoclonal antibodies resulting from the immortalisation of lymphocytes by Epstein Barr virus, and guidelines have been produced dealing with the quality aspects of such materials, which however have not been widely licensed. Much of the approach to virological testing of cells where there is little experience of their use in production or the clinic, including those intended for cell therapy, seems to derive from this type of product.

The approach is generally to establish a cell banking system where a master cell bank is used to prepare a working cell bank, which in turn is used to generate a production cell culture. The cell banks may be exhaustively tested for contamination and other properties over a period. Generally such cell banks are only acceptable if they are virus-free, although in some cases involving rodent lines the presence of non- infectious retroviruses may be acceptable.

Transmissions of Viral Infections by Biological Products and Appropriate Testing of Human Source Materials

Transmissions of serious disease from biological materials have occurred, particularly in the field of plasma-derived proteins, which have transmitted hepatitis A, B and C and HIV, as well as less serious viruses such as parvovirus B19. Transmission of hepatitis and HIV was particularly common about twenty years ago before current concerns about viruses were recognised and met. Fatal transmissions of Creutzfeld- Jakob Disease occurred when hormones or dura mater from cadaveric donors with the disease were used. There is no doubt of the seriousness of such events and they have informed the approaches to the safety of materials derived from human sources to a major extent. However these types of agent may be inappropriate causes of concern for other materials. Urokinase produced from cells cultured from human foreskin was found to be contaminated with enteric viruses including reo viruses on a number of occasions. Neither the common blood-borne viruses nor the agent of CJD are known to grow in cell culture, and the only reason for seeking them is that the starting cultures may be contaminated with blood; it is unlikely that they will be contaminated with the agent of CJD which is found in highest concentrations in brain from elderly individuals.

Where complex culture conditions are used, for example in attempts to generate artificial skin where cells of one type are cultivated on cells of another, viruses not found readily in either cell type may be activated. It is unlikely that any of them will be the common and hazardous blood-borne viruses which are difficult or impossible to cultivate.

An example of testing which has proved to be appropriate comes from the human diploid cell line MRC5, which was derived from termination of a pregnancy in 1968. The cells were derived from the lungs of the foetus and have been used extensively for the production of vaccines such as those against rubella, varicella, rabies and polio. They were studied in many laboratories for karyotypic stability and tumorigenesis, and for the presence of adventitious agents by methods which were state-of-the-art at the time. This meant chiefly the use of co-cultivation with different cell types for signs of transmissible agents giving cytopathic effects. They are free of agents detectable by these methods and have been safely used for many years in the production of a range of products, most of which, like varicella vaccines, are unpurified. While the origin of the cells pre-dates assays for any blood-borne virus, even hepatitis B, no blood-borne virus has ever been reported in them and it seems most unlikely that any such virus is present.

Testing of novel human cell lines and cells intended for cell therapies has tended to focus on viruses such as hepatitis B, C and HIV, the herpes viruses such as Epstein Barr virus, cytomegalovirus and T and B cell tropic viruses such as HTLV. It is possible that such viruses could be present, if only as contaminants from the blood. However, there are more likely contaminants depending on the source of the cells. As one example, where human foreskin cells are used, contamination by enteric viruses such as reo-, adeno-, corona- or entero-viruses may occur because of the location of the starting tissue. Equally, contamination by viruses found in urine may occur, including the polyoma viruses JC and BK, and cytomegaloviruses, some of which may not produce an aggressive cytopathic effect. Where a cell banking system is used, contamination by such 'relevant' viruses seems more of a risk than by blood- borne viruses. As the cells are to be used in cell therapies, the production process is unlikely to reduce or remove viral infectivity in a way comparable to that demonstrated for processes for the manufacture of fractionated blood proteins. Likewise the intimate contact between the patient and the living product makes the viral safety issues more pressing.

Summary and Conclusions

The testing of cells and starting materials for biological medicinal products made from human sources has tended to focus on blood-borne viruses and those with a tropism for B lymphocytes, probably arising from the existing guidance on human monoclonal antibodies and experience with proteins fractionated from human blood. This may not be the best course although blood may contaminate starting materials. Other agents may be more likely contaminants, and it is essential to apply sound virological principles to ensure product safety, particularly in instances such as cell therapy where the product is a living cell which is intended to persist in the recipient for some time.

Dr. P.D. Minor, N.I.B.S.C., Blanche Lane, South Mimms, Potters Bar, Herts. EN6 3QG, U.K.
E-mail: pminor@nibsc.ac.uk

Robertson J (ed): PDA/EMEA European Virus Safety Forum.
Dev Biol. Basel, Karger, 2004, vol 118, pp 155-163.

Animal Virus Contaminants of Biotechnology Products

D. Onions

BioReliance Inc, Glasgow, and University of Glasgow Department of Veterinary Pathology, U.K.

INTRODUCTION

Characterising potential animal virus contaminants in biotechnology products is a complex and constantly evolving problem. In general the perception has been that the nearer the product is to being sourced from an animal, the higher the potential risk so that xeno-transplantation is at the pinnacle of that risk profile, followed by transgenic systems. At the other end of the spectrum, well characterised, continuous cell lines are regarded as the safest production system, with a slightly higher risk being attached to primary cell production lines. While this might be an appropriate description of the potential hazards, risk also involves the probability of the hazard being realised. The highest risk profiles are therefore, associated with systems where there is little or no possibility of inactivating or removing adventitious agents. For instance, recombinant products produced from transgenic goats' milk that are subject to robust purification processes are inherently safer than live, unprocessed vaccines produced in primary cells using non-irradiated serum. However, although the risk of certain hazards may be very low, such as the possibility of TSE contamination of milk products, the consequences of that hazard may be so important that extraordinary reassurance may be needed on the safety of the production system.

Four broad, often overlapping, hazard groups of viruses can be identified with zoonotic viruses forming the leading group. The second group is formed from viruses which, although not zoonotic, may replicate in human cells in vitro. When introduced through parenteral routes or into immunocompromised persons, such viruses may establish productive infections in vivo. In the third group are animal DNA viruses that infect human cells but often undergo abortive replication, a process that may be accompanied by neoplastic transformation. Attention also needs to be paid to virus families that have displayed rapid changes in host range. For instance, canine parvovirus 2 was responsible for a devastating pandemic in the dog population in the late 1970s and early 1980s, yet this virus is believed to have originated from the cat (or closely related) parvovirus by a process involving as few as three mutations. The recent experience of HIV and SARS are testament to the evolving pattern of virus infections and emphasises that assumptions of safety should not rely solely on host-range characteristics. These categories of adventitious agent are discussed below, with examples from materials used in manufacture and cell substrates.

Zoonotic Contaminants from Foetal Bovine Serum

A surprisingly large number of manufacturing processes still use non-irradiated foetal calf serum (FCS) and, while this material is screened for bovine adventitious agents before use, it remains a potential source of contamination. This risk principally relates to the method of sourcing material in the United States from where most sera derive. Typically around one litre amounts are collected from third trimester foetuses and these are pooled into 100 litre sub-lots and then into larger pools of 1000 litres or more. Of necessity a 1000-litre pool may involve the collection from animals originating over a wide geographical area in the mid-west of the U.S. Best practice among manufacturers involves testing the sub-lots as well as the final serum pool. This procedure will reveal contamination by many of the critical adventitious agents but viruses with a low prevalence may escape detection.

Typical of these low prevalence contaminants are the bunyaviruses, particularly Cache Valley virus. The *Bunyaviridae* family contains over 300 viruses grouped into five genera. The genome of these viruses is segmented into three strands permitting complex reassortants to form, as occurs with influenza virus. Consequently, the taxonomy of these viruses can be complex. Bunyaviruses exhibit a wide in vitro host range and can often infect a diversity of mammalian species outside their main host. They are nearly all transmitted by mosquitoes and in some cases trans-ovarial transmission of the virus occurs within the vector. Within the bunyavirus genus the most relevant serogroups or genetic complexes are the Californian serogroup and the Bunyamwera serogroup.

The California serogroup contains 14 viruses that have a variety of reservoir species, and domestic animals may incidentally become infected. In a survey in Wisconsin, which involved both cattle and goats, seropositivity was less than 10% for La Crosse Virus, Trivittatus virus and Snowshoe hare virus. Jamestown virus was present in all species with a seroprevalence of 10% or above [1]. These viruses are more widely distributed than their names suggest, as viruses belonging to this California encephalitis complex have been demonstrated in sheep and people in Europe [2].

Within the Bunyamwera group, Cache valley virus (CVV) is a recognised pathogen of ruminants. Serological evidence indicates that CVV replicates in horses, dogs, pigs and human beings. The virus is widely distributed in the mid-west and Southern U.S.A. but it has been detected in many U.S. states [3,4]. In ruminants, in particular sheep, it is associated with teratogenic effects on the foetus [5]. Although no association has been found between this virus and neural tube defects in women, this possibility should be borne in mind when handling the virus [6]. Most human infections with CVV are probably sub-clinical although the virus has been implicated in serious disease, which eventually proved fatal [7]. The closely related Potosi virus, found in white-tailed deer, can also infect domestic ruminants. White tailed deer and other deer species may act as an important reservoir for both Potosi and Cache Valley viruses [4].

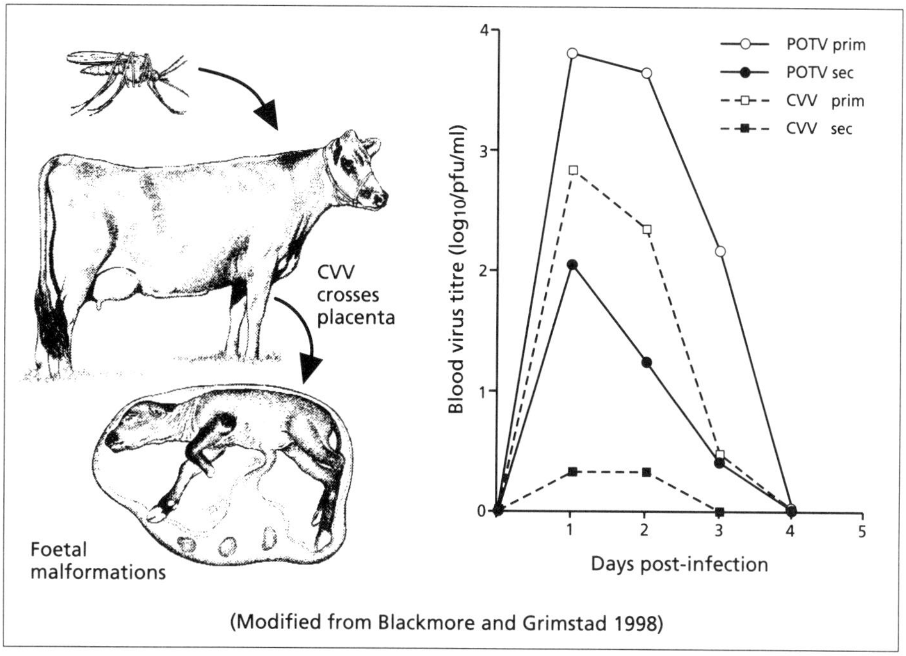

Fig. 1: Cache valley virus crosses the placenta and contaminates FBS.

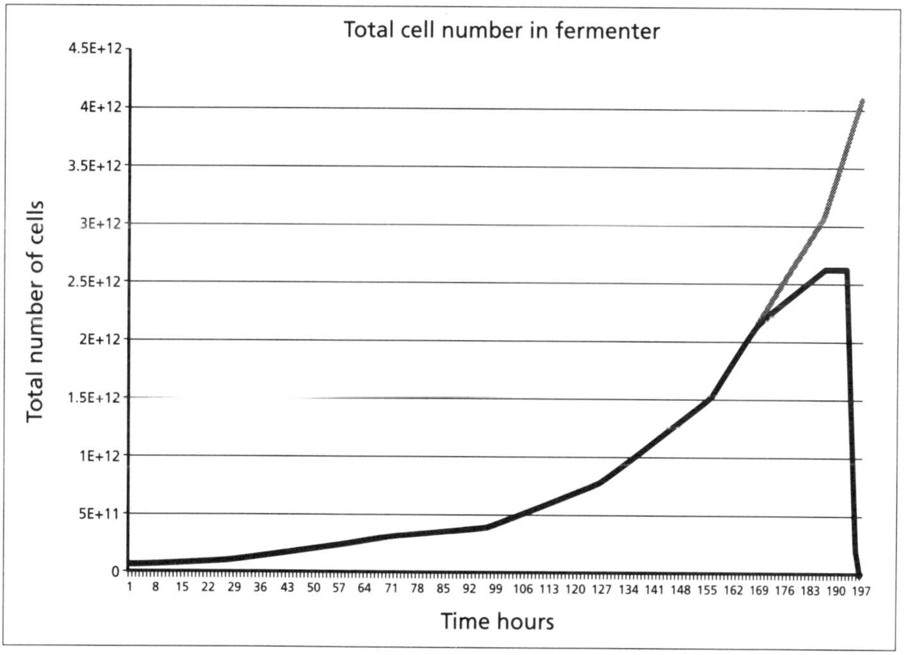

Fig. 2: Modelled Cell Population Dynamics in a fermenter following Introduction of 1 to 100 Infectious Units of Cache Valley Virus. Solid line uninfected fermenter, hatched line infected.

Adult cattle infected by CVV or Potosi virus (POTV) exhibit short-lived, one to three day, viraemias peaking at around 4 $\log_{10}$ i.u./ml (Fig. 1 & [ref 4]). During this period the virus may be passed to the foetus. Consequently, contamination of a serum pool is likely to be limited to a single donation and the titre of CVV in the pool is likely to be less than ~0.01 i.u./ml, a level which may be missed in routine testing.

In the past three years we have observed three major incidents involving the contamination of production systems with CVV. In each case they were costly in down-time and replacement of materials in the purification train. In one of these events, which involved a catastrophic crash in cell numbers in a fermenter on days seven to eight, we conducted experiments to determine the one-step growth curve for the contaminating CVV. It exhibited an eight-hour eclipse period followed by period of virus release and cell lysis that was complete by 30 hours. Based on these data and on an approximate cell population doubling time of 30 hours, we modelled the effects on cell number under conditions where only one infectious virus was present in the 20 litres of FBS employed in production. The modelled results indicated that, as observed, a sudden catastrophic fall in cell numbers would be expected at around day eight. Sensitivity analysis demonstrated that the results were independent of initial virus input over about 2 $\log_{10}$. Within the fermenter there was a high particle-to-infectivity ratio that indicated RT-PCR would have been a useful tool in detecting infection in the fermenter before any gross effects on cell numbers had occurred. Table 1 indicates the probability of detecting the virus using infectivity versus RT-PCR under the conditions observed in this fermenter crash.

Table 1: Probability of failure in detecting CVV contamination in a fermenter.

Hours post infection	Probability of negative infectivity result	Probability of negative RT-PCR result
60	1.00	0.16
72	0.97	0.00
84	0.83	0.00
96	0.34	0.00

While each of these contaminations was detected by appropriate tests on bulk harvests, the position might have been different if the production involved a lytic vaccine virus where a low but hazardous level of CVV might have remained. The most appropriate solution is to use irradiated serum in line with the proposals of the recent European guideline [8] but, in addition, RT-PCR testing of vaccine bulks would provide added security against contamination of vaccines by this potentially serious zoonotic virus.

Non-zoonotic Viruses that Replicate in Human Cells

Circoviruses are an emerging area of concern in biotechnology; they have the smallest physical (<17nm) and genome size of any mammalian virus so they are challenging to remove by filtration and have a very high resistance to gamma-irradiation and other physical inactivating steps [9,10]. This concern is heightened because of the potential cross species transmission of these viruses.

Until recently porcine circoviruses were the only known mammalian viruses in this family. Porcine circovirus (PCV) 1 is ubiquitous in pig populations but has not been associated with any syndrome. In contrast, PCV-2 is thought to play a part in the aetiology of post-weaning multi-systemic syndrome, one of the most important diseases affecting commercial pig production. Porcine circoviruses are not cytopathic in vitro and can rapidly establish persistent infections. Indeed this has hampered the development of infectivity assays, as nearly all stocks of the PK15 cells, the most sensitive cell line used for their detection, are contaminated by the virus. Recently, using a circovirus-free clone of PK15 cells, we have established an infectivity assay for the virus that uses a PCR endpoint. In our experience, nearly all commercial porcine trypsin sources contain detectable PCV genomes, so establishing infectivity assays as important in evaluating the status of this material.

Tischer et al [11] suggested that PCV may have neoplastic transforming properties based on the longer survival of persistently infected primary cells compared to mock infected cells, and on the ability of the former cells to form colonies in soft agar. However, there is no in vivo data to support an oncogenic role for these viruses, although PCV does replicate in mice in vivo and produces lesions [12]. PCVs can infect a range of other cells in vitro. Allan et al [13] demonstrated that PCV could replicate in primary leucocyte cultures of cattle but not human cells, while Fenaux et al (2000) reported replication in bovine and ovine cells as well as in Vero and HeLa cells. We have not found evidence of PCV infection in human or Vero master cell banks examined with a sensitive, quantitative, PCR assay but the ubiquity of the genomes in trypsin and the data on replication in human and Vero cells emphasise the importance of screening cells for this virus group.

Circoviruses and circovirus-related viruses have now been identified in other species. A novel virus with similar properties to a circovirus was identified by representational difference analysis in a human patient with non-A non-B hepatitis [14]. Subsequently it has become apparent that the majority of the human population is persistently infected by TTV-related viruses but that there is considerable genetic variability among isolates. Before the discovery of TTV, serological evidence of infection of humans, cattle and mice by a virus related to porcine circovirus had been presented [15]. Other surveys have not confirmed the presence of antibody to porcine circovirus 2 in cattle and humans [16,17]. However, a new circovirus has been isolated from cattle and the complete viral sequence obtained [18,19]. Similarly circoviral sequences have been identified in dogs (Nicolson and Onions, unpublished data) suggesting that these viruses are likely to be widely distributed in mammals. Given that new vaccine substrates based on canine, Syrian hamster and other cells are being developed, redundant PCR and other methods may be necessary to detect persistent infection by new members of this family.

Viruses with Potential Transforming Activity in Human Cells

Some DNA viruses are known to be oncogenic in their natural hosts and other species. These include the *Gamma-Herpesvirinae* and members of the *Polyomaviridae* and *Papillomaviridae*. Other DNA viruses, such as adenoviruses and certain poxviruses, may be oncogenic in heterologous hosts even when there is no oncogenic association in their species of origin. A common theme in these transformation events is that they are associated with abortive replication and usually with the integration of viral sequences into chromosomal DNA.

Between 1955 and 1963 poliovirus produced in monkey cells became contaminated by the polyomavirus, *Simian Virus 40* (SV40) and prompted concerns for the long- term health of the vaccine recipients, since SV40 is capable of inducing a range of tumours in hamsters including mesotheliomas and osteosarcomas [20]. Some early reports indicated the presence of SV40 T antigen in human ependymomas [21] but it was the advent of PCR analysis that raised major concern that SV40 sequences were present in mesotheliomas [22-25], osteosarcomas [26,27], and brain tumours [28,29]. Although several laboratories have reported similar results, the data have been controversial. For instance, some groups have failed to detect SV40 sequences in human tumours or have shown variable results between collaborating laboratories [30-33] but in the case of the Finnish data this has been attributed to non-use of contaminated vaccine [34]. The situation is even more polarised in the case of lymphomas where claims have been made that up to 40% of lymphomas may contain SV40 sequences [35,36], whereas others using highly sensitive quantitative PCR analysis have found uniformly negative results [37]. Moreover, even in those lymphomas where SV40 sequences have been detected, they are not clonally integrated, which raises questions about the aetiological significance of the observations.

The matter has been complicated further by the occurrence of SV40 sequences in tumours arising in patients that could not have received contaminated vaccine. The possibility that SV40 or a closely related virus are circulating in the human population has been addressed by serology. Initial studies indicated SV40 antibodies were present in a variable proportion of the population [38,39]. Additional analysis suggests that these results may be due to SV40 cross-reactive antibodies that develop in a proportion of people exposed to the known human polyomaviruses JCV and BKV [40,41]. Consequently there are still open questions on the presence and role of SV40 in human tumours and attempts to find a consensus view among workers in the field have not resolved all the issues [42,43].

SV40 is not the only polyomavirus of concern. Most batches of foetal calf serum (FCS) contain detectable sequences of bovine polyomavirus (BPyV) [44], a virus that is probably zoonotic, capable of replicating in some primate cells and which, like other polyomaviruses, has oncogenic properties [45-47]. The ubiquitous viral sequences in FCS have led to calls for the development of reliable assays to assess whether these sequences are associated with infectious virus [8]. A new infectivity assay with a PCR endpoint has been developed and will assist in ensuring the safety of FCS [48, D. Galbraith this volume].

Polyomaviruses may also be associated with novel cell substrates. Syrian hamsters can be infected by a polyomavirus that replicates in the hair follicle and can induce both epitheliomas and lymphomas [49]. Similarly, African green monkeys, the group of species from which Vero cells were originated, harbour a lymphotropic polyomavirus [50]. Surprising little attention has been paid to this virus, particularly as redundant PCR techniques have detected related viruses in other primates. Consequently, in the development of new cell substrates, we should be considering using redundant PCR systems and in vivo infectivity systems to detect novel polyomaviruses [51,52,37].

REFERENCES

1 Godsey MS Jr, Amoo F, Yuill TM, Defoliart GR: California serogroup virus infections in Wisconsin domestic animals. Am J Trop Med Hyg 1988 Oct;39(4):409-416.

2 Traavik T, Mehl R, Wiger R: Mosquito-borne arboviruses in Norway: further isolations and detection of antibodies to California encephalitis viruses in human, sheep and wildlife sera. J Hyg (Lond) 1985 Feb;94(1):111-122.

3 Calisher CH, Francy DB, Smith GC, Muth DJ, Lazuick JS, Karabatsos N, Jakob WL, McLean RG: Distribution of Bunyamwera serogroup viruses in North America, 1956-1984. Am J Trop Med Hyg 1986 Mar;35(2):429-443.

4 Blackmore CG, Grimstad PR: Cache Valley and Potosi viruses (Bunyaviridae) in white-tailed deer (Odocoileus virginianus): experimental infections and antibody prevalence in natural populations. Am J Trop Med Hyg 1998 Nov;59(5):704-709.

5 Edwards JF, Karabatsos N, Collinson EW, de la Concha Bermejilllo A: Ovine fetal malformations induced by in utero inoculation with Main Drain, San Angelo and Lacrosse viruses. Am J Trop Hyg 1997;56:171-176.

6 Edwards JF, Hendricks K: Lack of serological evidence for an association between Cache valley virus infection and anencephaly and other neural tube defects in Texas. Emerging Infectious Diseases 1997;3:195-197.

7 Sexton DJ, Rollin PE, Breitschwerdt EB, Corey GR, Myers SA, Dumais MR, Bowen MD, Goldsmith CS, Zaki SR, Nichol ST, Peters CJ, Ksiazek TG: Life-threatening Cache Valley virus infection. N Engl J Med 1997 Feb 20;336(8):547-549.

8 The European Agency for the Evaluation of Medicinal Products, Evaluation for Human Use, Note for Guidance on the Use of Bovine Serum in the Manufacture of Human Biological Medicinal Products (CPMP/BWP/1793/02) 25th April 2002.

9 Allan GM, Phenix KV, Todd D, McNulty MS: Some biological and physico-chemical properties of porcine circovirus. Zentralbl Veterinarmed [B] 1994 Mar;41(1):17-26.

10 Plavsic ZM, Bolin S: Resistance of porcine circovirus to gamma irradiation. BioPharm 2001 April:32-36.

11 Tischer I, Peters D, Pociuli S: Occurrence and role of an early antigen and evidence for transforming ability of porcine circovirus. Arch Virol 1995;140(10):1799-1816.

12 Kiupel M, Stevenson GW, Choi J, Latimer KS, Kanitz CL, Mittal SK: Viral replication and lesions in BALB/c mice experimentally inoculated with porcine circovirus isolated from a pig with postweaning multisystemic wasting disease. Vet Pathol 2001 Jan;38(1):74-82.

13 Allan GM, McNeilly F, Foster JC, Adair BM: Infection of leucocyte cell cultures derived from different species with pig circovirus. Vet Microbiol 1994 Aug 1;41(3):267-279.

14 Nishizawa T, Okamoto H, Konishi K, Yoshizawa H, Miyakawa Y, Mayumi M: A novel DNA virus (TTV) associated with elevated transaminase levels in posttransfusion hepatitis of unknown etiology. Biochem Biophys Res Commun. 1997 Dec 8;241(1):92-97.

15 Tischer I, Bode L, Apodaca J, Timm H, Peters D, Rasch R, Pociuli S, Gerike E: Presence of antibodies reacting with porcine circovirus in sera of humans, mice, and cattle. Arch Virol 1995;140(8):1427-1439.

16 Allan GM, McNeilly F, McNair I, Curran MD, Walker I, Ellis J, Konoby C, Kennedy S, Meehan B: Absence of evidence for porcine circovirus type 2 in cattle and humans, and lack of seroconversion or lesions in experimentally infected sheep. Arch Virol 2000;145(4):853-857.

17 Ellis JA, Konoby C, West KH, Allan GM, Krakowka S, McNeilly F, Meehan B, Walker I: Lack of antibodies to porcine circovirus type 2 virus in beef and dairy cattle and horses in western Canada. Can Vet J 2001 Jun;42(6):461-464.

18 Nayar GP, Hamel AL, Lin L, Sachvie C, Grudeski E, Spearman G: Evidence for circovirus in cattle with respiratory disease and from aborted bovine fetuses. Can Vet J 1999 Apr;40(4):277-278.

19 Hamel AL, Nayar GPS: Bovine circovirus complete genome. NCBI accession number NC_002068. 1998.

20 Carbone M, Stach R, Di Resta I, Pass HI, Rizzo P: Simian virus 40 oncogenesis in hamsters, in Brown F, Lewis AM Jr (eds): Simian virus 40 (SV40): A Possible Human Polyomavirus. Dev Biol Stand. Basel, Karger, 1998, vol 94, pp 273-279.

21 Tabuchi K, Kirsch WM, Van Buskirk JJ: Immunocytochemical evidence of SV 40-related T antigen in two human brain tumours of ependymal origin. Acta Neurochir (Wien). 1978;43(3-4):239-249.

22 Carbone M, Pass HI, Rizzo P, Marinetti M, Di Muzio M, Mew DJ, Levine AS, Procopio A: Simian virus 40-like DNA sequences in human pleural mesothelioma. Oncogene. 1994 Jun;9(6):1781-1790.

23 Galateau-Salle F, Bidet P, Iwatsubo Y, Gennetay E, Renier A, Letourneux M, Pairon JC, Moritz S, Brochard P, Jaurand MC, Freymuth F: SV40-like DNA sequences in pleural mesothelioma, bronchopulmonary carcinoma, and non-malignant pulmonary diseases. Pathol. 1998 Mar;184(3):252-257.

24 Pass H, Rizzo P, Donington J, Wu P, Carbone M: Further validation of SV40-like DNA in human pleural mesotheliomas, in Brown F, Lewis AM Jr (eds): Simian Virus 40 (SV40): A Possible Human Polyomavirus. Dev Biol Stand. Basel, Karger, 1998, vol 94, pp 143-145.

25 Cristaudo A, Powers A, Vivaldi A, Foddis R, Guglielmi G, Gattini V, Buselli R, Sensales G, Ciancia E, Ottenga F: SV40 can be reproducibly detected in paraffin-embedded mesothelioma samples. Anticancer Res 2000 Mar-Apr;20(2A):895-898.

26 Carbone M, Rizzo P, Procopio A, Giuliano M, Pass HI, Gebhardt MC, Mangham C, Hansen M, Malkin DF, Bushart G, Pompetti F, Picci P, Levine AS, Bergsagel JD, Garcea RL: SV40-like sequences in human bone tumors. Oncogene 1996 Aug 1;13(3):527-535.

27 Heinsohn S, Scholz RB, Weber B, Wittenstein B, Werner M, Delling G, Kempf-Bielack B, Setlak P, Bielack S, Kabisch H: SV40 sequences in human osteosarcoma of German origin. Anticancer Res 2000 Nov-Dec;20(6B):4539-4545.

28 Bergsagel DJ, Finegold MJ, Butel JS, Kupsky WJ, Garcea RL: DNA sequences similar to those of simian virus 40 in ependymomas and choroid plexus tumors of childhood. N Engl J Med 1992 Apr 9;326(15):988-993.

29 Martini F, Lazzarin L, Iaccheri L, Vignocchi B, Finocchiaro G, Magnani I, Serra M, Scotlandi K, Barbanti-Brodano G, Tognon M: Different simian virus 40 genomic regions and sequences homologous with SV40 large T antigen in DNA of human brain and bone tumors and of leukocytes from blood donors. Cancer 2002 Feb 15;94(4):1037-1048.

30 Chauvin C, Suh M, Remy C, Benabid AL: Failure to detect viral genomic sequences of three viruses (herpes simplex, simian virus 40 and adenovirus) in human and rat brain tumors. Ital J Neurol Sci 1990 Aug;11(4):347-357.

31 Krainer M, Schenk T, Zielinski CC, Muller C: Failure to confirm presence of SV40 sequences in human tumours. Eur J Cancer 1995 Oct;31A(11):1893.

32 Mulatero C, Surentheran T, Breuer J, Rudd RM : Simian virus 40 and human pleural mesothelioma. Thorax 1999 Jan;54(1):60-61.

33 Strickler HD: International SV40 Working Group A multicenter evaluation of assays for detection of SV40 DNA and results in masked mesothelioma specimens. Cancer Epidemiol Biomarkers Prev 2001 May;10(5):523-532.

34 Hirvonen A, Mattson K, Karjalainen A, Ollikainen T, Tammilehto L, Hovi T, Vainio H, Pass HI, Di Resta I, Carbone M, Linnainmaa K: Simian virus 40 (SV40)-like DNA sequences not detectable in finnish mesothelioma patients not exposed to SV40-contaminated polio vaccines. Mol Carcinog 1999 Oct;26(2):93-99.

35 Vilchez RA, Lednicky JA, Halvorson SJ, White ZS, Kozinetz CA, Butel JS: Detection of polyomavirus simian virus 40 tumor antigen DNA in AIDS-related systemic non-Hodgkin lymphoma. J Acquir Immune Defic Syndr 2002 Feb 1;29(2):109-116.

36 Shivapurkar N, Harada K, Reddy J, Scheuermann RH, Xu Y, McKenna RW, Milchgrub S, Kroft SH, Feng Z, Gazdar AF: Presence of simian virus 40 DNA sequences in human lymphomas. Lancet 2002 Mar 9;359(9309):851-852.

37 MacKenzie J, Wilson KS, Perry J, Gallagher A, Jarrett RF: Association between simian virus 40 DNA and lymphoma in the United kingdom. J Natl Cancer Inst 2003 Jul 2;95(13):1001-1003.

38 Butel JS, Wong C, Vilchez RA, Szucs G, Domok I, Kriz B, Slonim D, Adam E: Detection of antibodies to polyomavirus SV40 in two central European countries. Cent Eur J Public Health 2003 Mar;11(1):3-8.

39 Minor P, Pipkin P, Jarzebek Z, Knowles W: Studies of neutralising antibodies to SV40 in human sera. J Med Virol 2003 Jul;70(3):490-495.

40 Carter JJ, Madeleine MM, Wipf GC, Garcea RL, Pipkin PA, Minor PD, Galloway DA: Lack of serologic evidence for prevalent simian virus 40 infection in humans. J Natl Cancer Inst 2003 Oct 15;95(20):1522-1530.

41 Viscidi RP, Rollison DE, Viscidi E, Clayman B, Rubalcaba E, Daniel R, Major EO, Shah KV: Serological cross-reactivities between antibodies to simian virus 40, BK virus, and JC virus assessed by virus-like-particle-based enzyme immunoassays. Clin Diagn Lab Immunol 2003 Mar;10(2):278-285.

42 Klein G, Powers A, Croce C: Association of SV40 with human tumors. Oncogene 2002 Feb 14;21(8):1141-1149.

43 Carbone M, Pass HI, Miele L, Bocchetta M: New developments about the association of SV40 with human mesothelioma. Oncogene 2003 Aug 11;22(33):5173-5180.

44 van der Noordaa J, Sol CJ, Schuurman R: Bovine polyomavirus, a frequent contaminant of calf sera, in Brown F, Cartwright T, Horaud F, Spieser JM (eds): Animal Sera, Animal Sera Derivatives and Substitutes Used in the Manufacture of Pharmaceuticals: Viral Safety and Regulatory Aspects. Dev Biol Stand. Basel, Karger, 1999, vol 99, pp 45-47.

45 Parry JV, Lucas MH, Richmond JE, Gardner SD: Evidence for a bovine origin of the polyomavirus detected in foetal rhesus monkey kidney cells, FRhK-4 and -6. Arch Virol 1983;78(3-4):151-165.

46 Parry JV, Gardner SD: Human exposure to bovine polyomavirus: a zoonosis? Arch Virol 1986;87(3-4):287-296.

47 Schuurman R, van Strien A, van Steenis B, van der Noordaa J, Sol C: Bovine polyomavirus, a cell-transforming virus with tumorigenic potential. J Gen Virol 1992 Nov;73 (Pt 11):2871-2878.

48 Nairn C, Lovatt A, Galbraith DN: Detection of infectious Bovine polyomavirus. Biologicals 2003 Dec;31(4):303-306.

49 Vogel F, Rhode K, Scherneck S, Bastien C, Delmas V, Feunteun J: The hamster papovavirus: evolutionary relationships with other polyomaviruses. Virology 1986 Oct 30;154(2):335-343.

50 zur Hausen H, Gissmann L: Lymphotropic papovaviruses isolated from African green monkey and human cells. Med Microbiol Immunol (Berl) 1979 Aug;167(3):137-153.

51 Volter C, zur Hausen H, Alber D, de Villiers EM : A broad spectrum PCR method for the detection of polyomaviruses and avoidance of contamination by cloning vectors, in Brown F, Lewis AM Jr (eds): Simian Virus 40 (SV40): A Possible Human Polyomavirus. Dev Biol Stand. Basel, Karger, 1998, vol 94, pp 137-142.

52 Krause PR: Conceptual and experimental approaches to address product safety issues raised by novel cell substrates, in Brown F, Lewis AM Jr, Peden K, Krause P (eds): Dev Biol. Basel, Karger, 2001, vol 106, pp 45-48; discussion 49-50.

Dr. D. Onions, BioReliance Inc, Todd Campus, West of Scotland Science Park, Glasgow, G20 0XA, Scotland, U.K.
E-mail: donions@bioreliance.com

Robertson J (ed): PDA/EMEA European Virus Safety Forum.
Dev Biol. Basel, Karger, 2004, vol 118, pp 165-168.

Transgenic Animals for the Production of Biological Medicines

P.D. Minor

NIBSC, Potters Bar, Herts., U.K.

BACKGROUND

Developments in biotechnology have opened up new methods for the production of biological molecules of medicinal and other interest. The original sources of such materials were tissues or fluids such as blood for the preparation of clotting factors, or pancreas for the production of insulin. The use of culture or fermentation methods provides a greater control over the process and its possible contaminants including infectious agents. When this is linked to the expression of a protein of interest by genetic manipulation in a well characterised micro-organism such as a strain of yeast or a bacterium, the quality of the material and the scale on which it can be produced are both potentially greatly improved. There exist a number of medicines which could not be used if they were not produced by expression systems, including human growth hormone, where possible contamination by the agent of CJD makes the naturally occurring material unacceptable, or erythropoietin, which occurs in such small amounts naturally as to make its extraction impractical on a suitable scale.

Other expression systems have been recognised or developed over the past decade or so, including insect-based systems such as baculovirus expression where the yield is potentially very high and transgenic animals, which could act as living fermenters. The attractions of such systems are the ease and efficiency with which they can produce the desired starting material for down-stream processing. They are novel in some of the issues they raise, and since the first guidance documents on the production of medicines by the use of transgenic animals some of the concerns have become clearer, the main point being that the system involves old-style production methods based on living whole animals, which are difficult to control, mixed with advanced biological methods for establishing the system in the first place. Much of the current concern therefore centres around the infectious agents which may be present.

Advantages of Production by Transgenic Animals

Transgenic animals may be produced by very precise manipulation of the genome. The procedures are most highly developed for mice, where it is possible to introduce genes by specific replacement of the indigenous gene, place genes under highly effective promoters, express them in specific tissues at specific stages of development or under conditions where the gene is induced by environmental influences such as diet in which tetracycline is, or is not, included. The most sophisticated methods are not routinely applied to larger transgenic animals more suitable for production because of lack of experimental systems, but expression of genes of interest in mammary glands, so that the product is found in the milk, is a standard approach. The raw material for processing is thus secreted in a fluid which does not require sacrifice of the animal and which is generally of high safety in so far as it is the primary food of the offspring. The technology associated with milk production in a variety of animals including goats, sheep and cows is amongst the oldest in civilisation. Finally the yields can be extremely high, amounting to several milligrammes per litre. Where animals bred for milk production are used, the outputs can be such as to meet needs which other methods cannot. For example, it is possible that a single transgenic cow could produce sufficient surface antigen in its milk to meet the entire global need for vaccine against hepatitis B.

Issues Associated with the Use of Transgenic Animals in Production of Biological Medicines

Despite the advantages of a potentially safe starting material with an extremely high content of the substance of interest, there are numerous issues to be considered. There is a conflict between the production of medicines by GMP and normal agricultural practice. Thus agricultural animals such as cows or sheep cannot be kept adequately under the kinds of conditions developed for laboratory animals such as mice. Specifically they require a sophisticated gut flora to digest their food. This leaves them uncontrollably exposed to infection by a number of agents. Milk is routinely contaminated to some extent with faecal material reflected in the presence of coliform bacteria, which would not be acceptable in most GMP production facilities.

In most cases production of the biological could be carried out on a commercial scale by other less contentious methods, such as yeast or bacterial expression systems. In the case of hepatitis B vaccine cited above, production involves yeast rather than transgenic animals, for instance. There may also be difficulty in expressing highly active pharmaceutical products, such as erythropoietin, where sufficient hormone may leak from the mammary gland into the blood stream to have serious adverse effects on the animal when lactation begins. However there are products which must be used on a very large scale if they are to be effective, including protein C and alpha 1 antitrypsin, which can be purified from plasma but in quantities which are hardly useful. The use of the technology has concentrated on these types of materials.

Public acceptance of transgenic technology is not assured, although in the current examples it is likely that the animals would be treated extremely well because of their high value, and probably better than most typical farm animals.

One specific matter is the occurrence of viruses, where the demands of pharmaceutical production and agriculture do not overlap completely. Viruses of agricultural interest are those affecting yield or threatening life; agents such as herpes or adenoviruses which may have only slight effects on health are generally not regarded as major concerns. This is not the usually held view in the regulation of biological medicines, where any contaminating virus is a potential threat.

Animals Used in the Production of Biological Medicines by Transgenic Technology

All animals, including humans, will be expected to have their own large catalogue of infectious agents. Documents in draft which may be informative with respect to transgenic animal production include those dealing with animal sera for medicinal purposes such as anti-lymphocyte activity following transplantation or as anti-venoms or the veterinary concerns on the use of animal sera elaborated by the European Pharmacopoeia.

Rabbits

Rabbits can be kept under specified pathogen-free conditions and are standard, if large, laboratory animals. They are therefore attractive from the point of view of the infectious disease risks of the production method. While the scale of production is less than with agricultural animals and the technology of milking less well established, it is nonetheless possible to obtain a viable yield. Rabbits are coprophagic, which has lead some to question whether they can be freed of enteric infectious agents such as rota- or-reo viruses. The range of viruses which may be encountered includes rota-, reo-, papilloma-, pox-, herpes-, borna-, paramyxo- and adeno-viruses. The significance of these agents for human health is not clear largely because of lack of human exposure.

Pigs

Transgenic pigs have been proposed as a source of organs for human transplants where the contact between the animal tissue and the patient is very close and prolonged. The infectious disease risks have therefore been exhaustively considered, including virological and other agents [1]. Pigs can be kept under very clean conditions permanently indoors, and while it is difficult to prevent infection with cytomegaloviruses and the endogenous retroviruses of pigs, these do not necessarily pose the major threat to human health for a product which may be purified and given over a limited period of time and dosage. Milking of pigs is not a routine agricultural practice and poses some technical difficulties. Viruses which may be of concern include pseudorabies

Sheep

Sheep are used as agricultural sources of milk and the technology is well understood and developed. Unlike pigs or rabbits, it is generally accepted that they do not thrive when kept in containment and that as ruminants they require a sophisticated gut flora. They are subject to a number of viral infections, some of which can transmit to humans, as well as some which have been poorly studied because of their small agricultural or zoonotic significance. They include foot-and-mouth disease virus herpes and adenoviruses, poxviruses, flaviviruses including Louping Ill, a close relative of tick-borne encephalitis, re-, born-, paramyx- and retro-viruses, including Maedi-Visna virus, the archetypal lentivirus and scrapie, the archetypal TSE.

Cows

Cows are the primary milk-producing animals and have been bred to yield massive quantities. Like sheep they cannot be readily maintained in contained environment and require a gut flora to survive. Viruses of concern include foot-and-mouth disease virus, herpes-viruses, reo-, rota- and entero-viruses

Dealing with Infectious Agent Risks

Infectious risks presented by transgenic animal production may be approached in a way similar to that used for products fractionated from human plasma, by screening the animals for viruses of concern and testing the ability of the production process to remove or inactivate model viruses following principles established over many years. Agricultural animals, or those kept under agricultural conditions, are most unlikely to be virus-free, although they can clearly be kept free of the agents which pose the most significant threats to either human or their own health. In the instance of sheep kept in the open fields, it would be possible to maintain them free of Louping Ill and of the pox virus orf; in fact this was generally done in the days when sheep were used for the production of smallpox vaccines. It is unlikely that they will be kept free of adenoviruses or sheep herpesviruses, which are clinically silent. Where pigs were considered as donors of organs for transplantation, it proved possible to maintain them free of a surprisingly comprehensive range of viruses by keeping them under containment. This is considered impractical for the ruminants.

It may be useful to infect an animal deliberately with an agent and assess its presence in the milk under natural circumstances to provide assurance that even were the animal infected it would not pose a hazard.

This does not resolve the problem of whether any viruses can be safely ignored. Over the years the approach to human blood-borne viruses has become increasingly stringent and it is debatable if the known presence of any virus in starting material of human origin can be considered acceptable. Parvovirus B19 is regarded as a trivial childhood infection except in certain well-defined risk patients, but a substantial effort is currently going into minimising its presence in human blood products. On the other hand total exclusion is not being demanded because of the wide prevalence of the virus in plasma pools. This may offer a precedent for the transgenic animal concerns.

The virological matters are a serious obstacle to the general use of transgenic animals for production purposes, and the justification for their use will involve a risk-benefit analysis which is likely to include considering the possibility of other means of production.

Reference

1 Onions D, Cooper DK, Alexander TJ, Brown C, Claassen E, Foweraker JE, Harris DL, Mahy BW, Minor PD, Osterhaus AD, Pastoret PP, Yamanouchi K: An approach to the control of disease transmission in pig-to-human xenotransplantation. Xenotransplantation 2000;7:143-155.

Dr. P.D. Minor, N.I.B.S.C., Blanche Lane, South Mimms, Potters Bar, Herts. EN6 3QG, U.K.
E-mail: pminor@nibsc.ac.uk

List of Participants

ABUJOUB Amin A., Biogen Inc., 14 Cambridge Center, Cambridge, MA 02142-1453, USA.

ALLIGER Peter, Sandoz GmbH, Biochemiestrasse 10, AT-6250 Kundl.

AMIRPOUR Janne, Novo Nordisk A/S, Nybrovej 80, DK-2820 Gentofte.

ANDRAE Marie-Louise, Amersham Biosciences, Bjorkgatan 30, SE-751184 Uppsala.

ARMSTRONG Alison, BioReliance, Stirling University Innovation Park, Helfoots Road, Stirling FK9 4NF, United Kingdom.

ATWILER G., Navigant Biotechnololgies, 1215 Quail Street, Lakewood, CO 80215, USA.

BALL Philipp E., Novartis Pharma S.A.S., Centre de Biotechologie, 26 rue de la Chapelle, B.P. 349, FR-68333 Huningue.

BAR Liliana, Omrix Biopharmaceutical Ltd., Weizman Science Park Bldg. 14, P.O. Box 619, Ness Ziona 76106, Israel.

BAR Raphael, Pharmos Ltd., Bulding 13B, Kiryat Weizmann Pharma, Rehovot 76326, Israel.

BATE Anita, Eden Biopharm Ltd, D5, Stanlaw Abbey Business Centre, Dover Drive, Ellesmere Port, CH65 9BF, United Kingdom.

BAUER Ralph, Trioh Pharma GmbH, Frankfurtes Ring 193a, DE-80807 Munich.

BAUERFEIND Ursula, Blutspendedienst, Eldagsener Strasse 38, DE-31830 Springe.

BEAL Ernestine, Laboratoires Serono SA, Zone Industrielle, CH-1267 Coinsins.

BERESFORD Roland G., Navigant Biotechnologies, 1 Kerosken 28, BE-1930 Zaventem.

BIESERT Lothar, Octapharma Pharmaceutica, Paul-Ehrlich-Str. 42-44, DE-60596 Frankfurt.

BIRCH David W., Covance Laboratories Ltd., Otley Road, Harrogate HG3 1PY, United Kingdom.

BLOCK Wolfgang, Bayer AG, PH OP QA, E28, DE-51368 Leverkusen.

BLUEMEL Johannes, Paul-Ehrlich-Institut, Virus Safety Section, Paul-Ehrlich-Str. 51-59, DE-63225 Langen.

BLUSCH Jurgen, Novartis Pharma AG, W41-681.3.42, CH-4002 Basel.

BONAM Duane, Amgen Inc., 4000 Nelson Rd., Longmont, CO 80503-9004, USA.

BOSCHETTI Nicola, ZLB Bioplasma, Wankdorfstrase 10, CH-3000 Bern 22.

BRANDT Joan, MedImmune Inc., 660 Research Dr., Frederick, MD 21703, USA.

BRATBERG Kimberly A., IDEC Pharmaceuticals Corp 11011 Torreyana Rd., San Diego, CA 92191, USA.

BREITLING Gitte M., Danish Medicines Agency, Axel Heidesgade 1, DK-2300 Copenhagen S.

BREMNES Bjorn, Norwegian Medicines Agency, Sven Oftedals Vei 6, NO-0950 Oslo.

BREVER Beate, Biotest AG, Landsteinerstr. 5, DE-63303 Dreieich.

BRORSON Kurt A., FDA/CDER, NIH Bldg. 29B, Rm. 5E16, 29, Lincoln Dr., Bethesda, MD 20892, USA.

BROUMIS Connie, CSL Bioplasma, CSL Limited, 189-209 Camp Rd, Broadmeadows 3047, Australia.

BROWN Kevin, National Heart, Lung and Blood Institute, Building 10, Magnuson CC, Rm 7C218, 10 Center Dr., Bethesda, MD 20892, USA.

BRUCE Lynn, Wyeth BioPharma, Wyeth Media Ireland, Grange Castle International, Business Park, Dublin 22, Ireland.

BUCHHOLZ Christian J., Paul-Ehrlich-Institut, Paul-Ehrlich Str. 53-59, DE-63225 Langen.

BURNOUF Thierry, Human Plasma Product Services (HPPS), 18 Rue Saint-Jacques, FR-59800 Lille.

CARO Noemi, Chiron S.R.L., Via Fiorentina, 1, IT-53100 Siena.

CARTER Jeffrey R., Millipore Corporation, 80 Ashby Rd., Bedford, MA 01730-9125, USA.

CARTWRIGHT R., Ipsen BioPharma Ltd., Ash Rd., Wrexham Ind. Estate, Wrexham LL13 9UF, United Kingdom.

CASEY Barbara, ThromboGenics Ltd., 14 Bridgecourt Office Park, Walkinstown Ave., Dublin 12, Ireland.

CELIS Patrick, EMEA, 7 Westferry Circus, Canary Wharf, London E14 4HB, United Kingdom.

CHANG Jen-Ting, Animal Technology Institute, Miaoli, Chu-Nan, P.O. Box 23, R.O.C Taiwan.

CHEN Qi, Genentech, Inc., Rm. 20016 Dept of Recovery Sci., 1 DNA Way M/S 70, South San Francisco, CA 94080, USA.

CHOU Yu-Chi, Animal Technology Institute, Miaoli, Chu-Nan, P.O. Box 23, R.O.C. Taiwan.

CROSS Philip J., Ste. 407, 4 Blackfan Circle, Boston, MA 02115, USA.

CUPERUS Roelck A., Genzyme Europe BV Gooimeer 10, NL-1411 DD Naarden.

DANI Maria, Tecnogen S.p.A., Loc. La Fagianeria, IT-81015 Piana di Monte Verna.

DEAN Melanie, Millipore Corporation, 80 Ashby Rd., Bedford, MA 01730, USA.

DEKKER Robert J, RIVM, A. Van Leeuwenhoeldaan 9, Gebouiweg 12, NL-3721 MA Bilthoven.

DIBELLA Elsie E., Centocor Inc., 200 Great Valley Pkwy, Malvern, PA 19355, USA.

DICHTELMUELLER Herbert, Biotest Pharma GmbH, Landsteinerstrasse 5, DE-63303 Dreieich.

DODT Johannes, Paul-Ehrlich-Institut, Paul-Ehrlich Str. 53-59, DE-63225 Langen.

DOMINGO Silvia, EMEA, 7 Westerferry Circus, Canary Wharf, London E14 4HB, United Kingdom.

DOODKORTE Caroline, Centocor, Inc., 200 Great Valley Parkway, Malvern, PA 19355, USA.

DRIESPRONG Paul, Centocor Europe BV, P.O. Box 251, Einsteinweg 101, NL-2300 AG Leiden.

DROHAN William N., Clearant Inc., 2nd Floor, 401 Professional Dr., Gaithersburg, MD 20879, USA.

DUHRKOP Mikkel, HemeBiotech A/S, Roskildevej 12C, DK-3400 Hillerod.

EDGINGTON Michael, MicroSafe BV, Niels Bohrweg 11-13, NL-2333 CA Leiden.

EDWARDS Treena, MedImmune UK Ltd., Plot 6, Renaissance Way, Blvd. Industrial Park, Liverpool L24 9JW, United Kingdom.

ERDMANN Carsten, Cuno, Wilhelm Theodor Romheld Str. 32, DE-55130 Mainz.

ESKLING Marie I., Pharmexa, Kogle Alle' 6, DK-2970 Horsholm.

EVERS Theo, European Plasma Fractionation Assoc., Plesmanlaan 125, NL-1066 CX Amsterdam.

FAGNANI Sabina, Kedrion SpA, Via Provinciale 1, IT-55027 Lucca Bolognana.

FARRUGIA Albert, Therapeutic Goods Administration, P.O. Box 100, Woden, ACT 2606, Australia.

FARSHID Mahmood, FDA, HFM-345 #200N, 1401 Rockvile Pike, Rockville, MD 20852-1448, USA.

FLAN Benoît, L F B, 3 av. des Tropiques, BP 305, FR-91958 Courtaboeuf Cedex.

FORTE Serene, Millipore Corporation, Mailstop B2D, 80 Ashby Rd., Bedford, MA 01730, USA.

FRINK Martin C, Solvay Pharmaceuticals, Hans-Boeckler-Allee 20, DE-30173 Hanover.

FUSE Akira, National Institute of Infectious Diseases, 4-7-1 Gakuen, Musashimurayama-Shi, Tokyo 208-0011, Japan.

GALBRAITH Daniel N., MedImmune Ltd, Blvd Industrial Estate, Speke, Liverpool L24 9JW, United Kingdom.

GARLICK Robert L, Pfizer Corporation, MS 1400-89-1, 7000 Portage Road, Kalamazoo, MI 49001, USA.

GEIGERT John, BioPharmaceutical Quality Solutions, 3533 Corte Esperanza, Carlsbad, CA 92009, USA.

GEORGE Karen, Centocor Inc., 200 Great Valley Parkway, Malvern, PA 19355, USA.

GEORGIADOU Evanthia, National Organization for Medicines, 284 Mesogion Avenue, Cholargos, GR-15562 Athens.

GERLICH Wolfram, Institut fuer Med. Virologie der Justus-Liebig-Universitaet, Frankfurter Strasse 107, DE-35392 Giessen.

GIJSENS Antoon, EMEA, 7 Westerferry Circus, Canary Wharf, London E14 4HB, United Kingdom.

GILGEN Michael, Swissmedic, Swiss Agency for Therapeutic Products, Erlachstr 3, CH-3000 Bern 9.

GOESTL Georg, Baxter AG, Industriestrasse 67, AT-1221 Vienna.

GRAF Ernst-Guenter, Pall Corporation GmbH, Philipp Reis Strasse 6, DE-63303 Dreieich.
GREGERSEN Jens P., Chiron Behring GmbH & Co., P.O. Box 1630, DE-35006 Marburg.
GROENER Albrecht, Aventis Behring GmbH, Emil-von-Behring-Strasse 76, DE-35041 Marburg.
GUENTHER Roland K, Novartis Pharma AG, P.O. Box, WSJ-87.2.12, CH-4002 Basel.

HAENEL Heinz H, Aventis Pharma Deutschland, H839, DE-65296 Frankfurt.
HARVEY Lorna, Inveresk Research, East Lothian, Tranent, EH22 3NE, Scotland, United Kingdom.
HEIDEN Magarethe, Paul-Ehrlich-Institut, Paul-Ehrlich Str. 53-59, DE-63225 Langen.
HEINTEL-REBSTOCK Dorothee, Paul-Ehrlich-Institut, Paul-Ehrlich Str. 53-59, DE-63225 Langen.
HEPBILDIKLER Stefan, Roche Pharmaceutical Biotech Production, Roche Diagnostics GmbH, Werk Penzberg, DE-82977 Penzberg.
HOEFNAGEL Marcel, RIVM, A. Van Leeuwenhoeldaan 9, NL-3721 MA Bilthoven.
HORN Hiltrud, Horn Pharmaceutical Cons., Am Alten Schloss 1, DE-76646 Bruchsal.
HUGHES Kimberly, EMD Pharmaceuticals, 45B Middlesex Turnpike, Billerica, MA 01821, USA.
HURD Louise, TKT UK Ltd., Wallington House, East Road, Cambridge CB1 1BH, United Kingdom.

JANDER Elisabeth, Pall Corp. GmbH, Philipp-Reis Str. 6, DE-63303 Dreieich.
JANSSEN Susan, RIVM, A. Van Leeuwenhoeldaan 9, NL-3721 MA Bilthoven.
JEHAN-KIMMEL Christine, LFB, BP 305, 3 av. des Tropiques (1-863EBS), FR-91958 Courtaboeuf Cedex.

KAERSGAARD Per, Novo Nordisk A/S, Hagedornsvej 1, DK-2820 Gentofte.
KAISER Klaus, Bayer AG, Wuppertal, Germany.
KAORU Yoshinari, Asahi Kasei Corp/ Planova Division, 9-1 Kanda Mitoshiro-cho, Chi Yoda-ku, Tokyo 101-8481, Japan.
KEDERER Elke, Asahi Kasei Planova Europe, Planova DIV./Grosse Telegraphenstr., DE-50676 Kolen.
KELLINGS Klaus P., Newlab Bioquality AG, Max-Planck-Str. 15A, DE-40699 Erkrath.
KERN Gerd, Millipore S.A., 39, rue de la Hardt, FR-67120-Molsheim.
KIERAN Jane E., Chiron, M/S H-100, 4560 Horton St., Emeryville, CA 94608, USA.
KIONKA Christine, E. Hoffmann - La Roche, Grenzacher Str, CH-4070 Basel.
KLEIN Ralf, BioReliance, Moningerstr. 12, DE-76135 Karlsruhe.
KNAPP Stefan, Chiron Behring GmbH & Co, Emil-von-Behring Strasse, DE-35041 Marburg.
KORHOLA Paula, National Agency for Medicine, P.O. Box 55, FI-00301 Helsinki.
KRAFFT Bianca, Paion GmbH, Martinstrasse 10-12, DE-52062 Aachen.
KREIL Thomas R., BioScience Baxter AG, Benatzkyg. 2-6, AT-1220 Vienna.
KRONENBERG Anke, Newlab BioQuality AG, Max-Planck-Str 15A, DE-40699 Erkrath.
KUSHIDA Katsutoshi, Wako Pure Chemical Industries Ltd., BioMedical Equipment Sys. Dept Div, 1-2, Doshomachi 3-Chome Chuo-Ku, Osaka 540-8605, Japan.

LAMMERS Frank, Aventis Pharma Deutschland GmbH, Quality Diabel GmbH Operations G680, Industriepark Hoechst, DE-65926 Frankfurt.
LARSEN Lars E, Novo Nordisk A/S, Moerkhoi Bygade 28, DK-2860 Soeborg.
LASKE Reiner, Aventis Behring GmbH, Emil Von Behring Str. 76, DE-35041 Marburg.
LAURSEN Inga A., Statens Serum Institute, Dept. of Research and Development, Artillerivej 5, DK-2300 Copenhagen.
LAUSCH Ralf, Sartorius AG, Postfach 32 43, Weender Landstrasse, 94-108, DE-37075 Göttingen.
LEE In Eui, Nihon Pall, Ltd., 46 Kasuminosato, Ami-machi, Inashiki-gun, Ibaraki 300-0315, Japan.
LEININGER Elizabeth, Biologics Consulting Group, 712 Dalehurst Avenue, San Mateo, CA 94403, USA.
LEUBE Joachim, Bayer Biologicals, Localita Bellaria di Rosia, IT-53018 Sovicille Si.
LIEBAUT Gilberte, Glaxo Smith Kline Biologicals, Rue de l'Institut 89, BE-1330 Rixensart.
LINNAVUORI Kimmo, National Agency for Medicines/Medical Devices, P.O.Box 55, Mannerheimintie 170, FI-00301Helsinki.
LLOPART Sylvie, Sanofi-Synthelabo Recherche, Labege Innopole-Voie 1, B.P. 137, FR-31676 Labège Cedex.

LOEWER Johannes, Paul Ehrlich Institute, Paul Ehrlich Str. 51-59, DE-63225 Langen.
LOH Wolfgang, Immune Biotec Pharma, Im Dammwald 27, DE-61381 Friedricksdorf.
LOVENSTEINA Inese, Millipore Corporation, 80 Ashby Road, Bedford, MA 01730, USA.
LUTZ Herbert, Millipore Corporation, 80 Ashby Rd., Bedford, MA 01730-2237, USA.
LYNE Katie, GlaxoSmithKline, South Eden Park Road, Beckenham, Berks. BR3 3BS, United Kingdom.

MacLEAN Christine, Q-One Biotech Ltd., Todd Campus, West of Scotland Science Park, Glasgow G20 OXA, United Kingdom.
MacLEOD Alex J, Scottish National Blood Transfusion Service, Protein Fractionation Centre, Ellen's Glen Road, Edinburgh EH17 7QT, Scotland, United Kingdom.
MacNAUGHTON Malcolm R., Inveresk Research, Tranent, EH33 2NE, Scotland,United Kingdom.
MACQ Anne F., GlaxoSmithKline Biologicals SA, Rue de l'Institut 89, BE-1330 Rixensart.
MAGATA Junji, Nihon Millipore K.K, Mita Kokusai Bldg. 15F, 4-28, Mita 1 chome Minato-Ku, Tokyo 108-0073, Japan.
MANNESSE Maurice, Pharming Technologies, Archimedesweg 4, NL-2333 CN Leiden.
MARKS Frank, DRK-Plasma Verarbeitungs Gesellschaft GmbH, Eldagsener Strasse 38, DE-31832 Springe.
MARTIN Carl N., Covance Laboratories, Otley Road, Harrogate, North Yorkshire HG3 1PY, United Kingdom.
MATHYS Esther, Scientific Institute of Public Health, J. Wytsmanstraat 14, BE-1050 Brussels.
MAZY Christopher, GlaxoSmithKline Biologicals SA, Bibliothèque, Rue de l'Institut 89, BE-1330 Rixensart.
McALISTER Morven B, Pall Corporation, 25 Harbor Park Dr., Port Washington, NY 11050-4630, USA.
McKENNEY Keith, Clearant Inc., 401 Professional Drive, Gaithersburg, MD 20879, USA.
MENDES Eva C., Infarmed, Parque de Saude de Lisboa, Av. do Brasil, 53, PT-1749-004 Lisboa.
MERTENS Hubert, Bayer AG, BHC-PH-OP-QA-IQM, Bldg. E28, DE-51368 Leverkusen.
MILLER Alex, Invitrogen, Ltd., 3 Fountain Dr., Inchinnan Bus. Park, Paisley PA4 9RF, United Kingdom.
MINOR Philip, N.I.B.S.C., Blanche Lane, South Mimms, Potters Bar, Herts EN6 3QG, United Kingdom.
MOERITZ Anne-Marie, Novartis Pharma AG, WKL - 121.211, CH-4002 Basel.
MORITA Toshi, Chugai Pharmaceuticals, 4-42-4 Honkomagome, Bunkyo-ku, Tokyo 113-0021, Japan.
MULKERRIN Michael, Abgenix Inc., 6701 Kaiser Dr., Mailstop 26, Fremont, CA 94555, USA.

NAIR Suresh, Ipsen Ltd., 190 Bath Road, Slough, Berks. SL1 3XE, United Kingdom.
NEUGEBAUER Brigitte, EMEA, 7 West Ferry Circus, Canary Wharf, London E14 4HB, United Kingdom.
NICOLAS Dumey, Texcell SA, 25 rue du Dr Roux, FR-75015 Paris.
NORLING Lenore, Genentech, Inc., 1 DNA Way, South San Francisco, CA 94080, USA.
NOWAK Thomas, Aventis Behring GmbH, P.O. Box 1230, DE-35002 Marburg.

O'CONNELL Barbara D., Wyeth BioPharma, One Burtt Rd, Andover, MA 01810, USA.
ODA Masahiro, Nihon Pall, Ltd., Oak Tower Building 17F, 6-8-1, Nishishinjuku, Shinjuku-ku, Tokyo 163-6017, Japan.
ONIONS David E, Q-One Biotech Ltd., Todd Campus, West of Scotland Science Park, Glasgow G20 0XA, United Kingdom.

PATEL Navin P., Invitorgen Limited, 3 Fountain Drive, Inchinnan Business Park, Renfrewshir PA4 9RF, United Kingdom.
PAULI Georg, Robert Koch Institute, Nordufer 20, DE-13353 Berlin.
PEIKER Marcus, Rentschler Biotechnologie GmbH & Co. KG, Erwin-Rentschler Strasse 21, DE-88471 Laupheim.
PELLOQUIN Francois, Aventis Pasteur, 1541 Avenue Marcel-Mérieux, FR-69280 Marcy L'Etoile.
PERMANNE Philippe, GlaxoSmithKline Biologicals SA, Rue de l'Institut 89, BE-1330 Rixensart.
PETERS Hermann, Rentschler Biotechnologie GmbH & Co. KG, Erwin-Rentschler Strasse 21, DE-88471 Laupheim.
PFEIFFER Gunther, Bayer AG, BHC-PH-OP-ELB-QC, Bldg. 302, DE-42096 Elberfeld.

PICKETT Andrew M., Ipsen Ltd., 190 Bath Road, Slough, Berkshire SL1 3XE, United Kingdom.

PIERARD Isabelle A., GlaxoSmithKline Biologicals SA, Rue de l'Institut 89, BE-1330 Rixensart.

PIFAT Dominique, Bayer Corporation, 85 TW Alexander Dr., Research Triangle Park, NC 27709, USA

PLANE Anthony B.L., Quara S.A., 6 Chemin des Glycines, CH-1170 Aubonne.

PORTELA Agustin, Agencia Espanola del Medicamento, Division de Productos Biologicos, Majadakonda 28220, ES-28220 Madrid.

POTTS Barbara, Genentech, Inc., MS #73, One DNA Way, S. San Francisco, CA 94080-4990, USA.

PREUSS Turid, Novo Norisk A/S, Mcerkhoej Bygagde 28 - SMH2.10.1, DK-2860 Soeborg.

PRIKHODKO Grigori G., American Red Cross Holland Laboratory, 15601 Crabbs Branch Way, Rockville, MD 20855, USA.

PURVES John, EMEA, 7 Westferry Circus, Canary Wharf, London E14 4HB, USA.

RAMDHANI Fahd, MicroSafe B.V., Niels Bohrweg, NL-2333 CA Leiden.

REMPETERS Gerold, DRK-Plasmaverarbeitungs GmbH, Eldagsenet Strasse 38, DE-31832 Springe.

ROBERTS Peter L., Bio Products Laboratory, Dagger Lane, Elstree, Hertsfordshire WDG 3BX, United Kingdom.

ROBERTSON James S, National Institute for Biological Standards and Control, Blanche Lane, South Mimms, Potters Bar, Herts ENG 3QG, United Kingdom.

ROESSLING Georg L., Schering AG, Muellerstr 178, DE-13342 Berlin.

ROTHWELL Liz, Wyeth Biopharma, Wyeth Medica Ireland, Grange Castle International, Business Park, Dublin 22, Ireland.

RUBINO Mike, Eli Lilly and Co, Lilly Corp Center, Bldg. DC 3832, 314, Indianapolis, IN 46285, USA.

RUFFING Michael, Boehringer Ingelheim Pharma GmbH & Co. KG, Birkendorfer Str. 65, DE-88397 Biberach.

SALAHEDDIME Mourad, Pharming Technologies, Archimedesweg 4, NL-2333 CH Leiden.

SALDANHA John, Canadian Blood Service, Head Office, 1800 Vista Alta, Ottawa, ON K1G 4J5, Canada.

SAMMANS Christine, Baxter, 6 Av. Louis Pasteur, FR-78311 Maurepas Cedex.

SCHAFER Werner, Amersham BioSciences Europe GmbH, Munzinger Strasse 9, DE-79111 Freiburg.

SCHEFER Peter, Berna Biotech Ltd, Rehhagstrasse 79, CH-3018 Berne.

SCHEIBLAUER Heiner, Paul-Ehrlich-Institut, Paul-Ehrlich Str. 53-59, DE-63225 Langen.

SCHMITT Klaus, Aventis Behring GmbH, P.O. Box 1230, DE-35002 Marburg.

SCHONNING Bolette, Novo Nordisk A/S, Morkhoj Bygade, 28, SMHS10, DK-2860 Soborg.

SCHUELER Eckhard, Aventis Behring GmbH, P.O. Box 1230, DE-35041 Marburg.

SCOTT Dorothy E., US Food & Drug Administration/CBER, Bldg. 29/Rm 232, 8800 Rockville Pike, Bethesda, MD 20892, USA.

SEEMANN Gerhard, Aventis Behring GmbH, P.O. Box 1230, DE-35002 Marburg.

SEIDEL Lutz, Medac GmbH, Theater Str. 6, DE-22880 Wedel.

SEITZ Rainer, Paul-Ehrlich-Institut, Paul-Ehrlich Str. 53-59, DE-63225 Langen.

SHANKS Michael R, Wellstat Biologics Corp., 45 W. Walkins Mill Rd., Gaithersburg, MD 20878, USA.

SILVESTER Glenda, EMEA, 7 Westferry Circus, Canary Wharf, London E14 4HB, United Kingdom.

SIMESEN Ruth B, Novo Nordisk A/S, Moerkhij Bygade 28, DK-2860 Soeborg.

SOBOLL Heidemarie, Paul-Ehrlich-Institut, Paul-Ehrlich Str. 53-59, DE-63225 Langen.

SOFER Gail, GE Healthcare, 1 Mason Hill Rd., Warren, NJ 20850-3349, USA.

STAACK Harold M, Abbott Laboratories, D-456, Bldg. R1B-6, 1401 Sheridan Rd., N. Chicago, IL 60064, USA.

STAGG Shirley, Bio Products Laboratory, Dagger Lane, Elstree, Hertfordshire W06 3BX, United Kingdom.

STARR-SPIRES Linda, Aventis Pasteur, Inc., Discovery Drive, Swiftwater, PA 18370, USA.

STEIB Alexandra, QLT Inc., 887 Great Northern Way, Vancouver, BC V5T 4T5, Canada.

STEINBORN Birgitta, Research & Development, Johnson & Johnson Pharmaceuticals, Hochstrasse 201, CH-8205 Schaffausen.

STUERTZ Kristin, Terring GmbH, Wittland 11, DE-24109 Kiel.

STUHLER Albert, Paul Ehrlich Institute, Viral Safety Section, Paul Ehrlich Str. 51 59, DE-63225 Langen.

SUNDLOF Jan, Carmetec AB, Kunesanesvagen 29, SE-75108 Uppsala.

SVENSSON Hakan, Xcyte Therapies Inc., 1124 Columbia, Suite 130, Seattle, WA 98104, USA.

SZENCZI Arpad, Human Pharmaceutical Works Co. Ltd., Tancsics M.U. 82, HU-2100 Godollo.

TANIMOTO Junjl, N.V. Asahi Vasel Planova Europe S.A., Paepseh Business Park, Paepsem 22, BE-1070 Brussels.

TARRACH Klaus, Sartorius AG, Weendea Landstrasse 94-108, DE-37075 Göttingen.

TERRAB Abdel, King Pharmaceuticals, Inc., 2232 Evergreen Road, Middleton, WI 53562, USA.

THUESEN Marianne H., CMC Biopharmaceuticals, Fruebjergvej 3, DK-2100 Copenhagen.

TOENJES Ralf, Paul-Ehrlich-Institut, Paul-Ehrlich Str. 53-59, DE-63225 Langen.

TOMINAGA Akihiro, Seikagaku Corporation, 1253 Tatemo, 3 Chome, Higashiyamato-Shi, Tokyo 207-0021, Japan.

TONGUE Suzanne, EMD Pharmaceuticals, 45B Middlesex Turnpike, Billerica, MA 01821, USA.

TRIEBSCH Oliver, Pall SeitzSchenk Filtersystems GmbH, Planiger StraBe 137, DE-55543 Bad Kreuznach.

TROUVIN Jean-Hugues, Agence Française de Sécurité Sanitaire des Produits de Santé, 143-147 Bd. Anatole France, FR-93285 Saint-Denis Cedex.

TURNER Judy C., Eli Lilly and Co, Lilly Technology Center, DC 3418, Indianapolis, IN 46285, USA.

UMINO Yukiko, National Institute of Infectious Diseases, 4-7-1 Gakuen, Musashimurayama-shi, Tokyo, 208-0011, Japan.

VAN ENGELENBURG Frank, CLB Sawquin Blood Supply Fondation, Plesmanlaan 125, NL-1066 CX Amsterdam.

VAN NOTE Karen M., Chiron Corporation, 4560 Horton St., H-100, Emeryville, CA 94608-2916, USA.

VAN SOMMEREN Ton, Diosynth BV, P O Box 20, Room 1035, NL-5340 BH Oss.

VICARI Giuseppe, Via Principessa Clotilde 5, IT-00196 Roma.

VON HOEGEN Ilka, PPTA, Boulevard Brand Whitlock 114/5, BE-1200 Brussels.

VYNCK Samuel, Asahi Kasei Planova Europe, Paepsem Business Park, Blv. Paepsem 22, BE-1070 Brussels.

WAFELMAN Amon, Leiden Universtiy Medical Ctr., P.O. Box 9600, NL-2300 Leiden.

WAINWRIGHT William, Wyeth Lederle Vaccines, 206 North Biddle St., P.O. Box 304, Marietta, PA 17547, USA.

WANG Shih-Rong, Animal Technology Institute, Miaoli, Chu-Nan, P.O. Box 23, R.O.C. Taiwan.

WAWOTZNY Ralf, Sartorius AG, Postfach 32 43, Weender Landstrasse 94-108, DE-37075 Göttingen.

WEBB Paul D., Ipsen BioPharm Ltd., Ash Road, Wrexham Ind. Estate, Wrexham LL13 9UF, United Kingdom.

WEINER Olaf, November AG, Department for Molecular Medicine, Ulrich-Schalk-Str. 3, DE-91056 Erlangen.

WENDE Martina-Aimee, BVL Bundesafm Verbraucherschertz, Diedersdorfer Weg 1, DE-12277 Berlin.

WESSTROM Christine, Biovitrum AB, Lindhagens Gatan 133, SE-11276 Stockholm.

WIESMANN Ralf, Paion GmbH, Martinstr., 10-12, DE-52062 Aachen.

WILLKOMMEN Hannelore, Clearant, 401 Professional Drive, Gaithersburg, MD 20979, USA.

WINTER Helmut, Boehringer Ingelheim, Birkendorfer Strasse 65, DE-88397 Biberach an der Riss.

WOENSEL Peter V., Intervet International BV Wim Korverstraat 35, , NL-5830 AA Boxmeer.

WOJCIECHOWSKI Peter W., Centocor Inc., 200 Great Valley Pkwy, Malvern, PA 19355-1307, USA.

WOLFF Gilbert, Aventis Pharma SA, Mail Stop, 20 Av. Raymond Aron, FR-92165.

WONNACOTT David, Inhibitex, Inc., 1165 Sanctuary Parkway, Suite 400, Alpharetta, GA 30004, USA.

XU Yuan, GlaxoSmithKline, 709 Swedeland Road, UE3839, King of Prussia, PA 19406, USA.

ZAPATA Gerardo, Abgenix Inc., MS 26, 6701 Kaiser Dr., Fremont, CA 94555, USA.

ZERLAUTH Gerold, Baxter AG, Industriestrasse 67, AT-1220 Vienna.

Index of Authors

Developments in Biological Standardization

Vol. 80: Heidelberg, 1992	Transmissible spongiform encephalopathies-Impact on animal and human health.
Vol. 81: Cannes, 1992	Virological safety aspects of plasma derivatives.
Vol. 82: Albany, NY, 1993	Recombinant vectors in vaccine development.
Vol. 83: Annecy, 1993	Genetic stability and recombinant product consistency.
Vol. 84: Langen, 1993	Non-target effects of live vaccines.
Vol. 85: Santa Fee, NM, 1994	Genetics of streptococci, enterococci and lactococci.
Vol. 86: Langen, 1994	Replacement, reduction and refinement of animal experiments in the development and control of biological products.
Vol. 87: Geneva, 1995	New approaches to stabilisation of vaccines potency.
Vol. 88: Bethesda, MD, 1995	Viral safety and evaluation of viral clearance from biopharmaceutical products.
Vol. 89: Rome, 1995	Pertussis vaccine trials.
Vol. 90: Oslo, 1996	Fish vaccinology.
Vol. 91: San Francisco, CA, 1996	Development of specifications for biotechnology pharmaceutical products.
Vol. 92: Bergen, 1996	Modulation of the immune response to vaccine antigens.
Vol. 93: Annecy, 1996	Safety of biological products prepared from mammalian cell culture.
Vol. 94: Bethesda, MD, 1997	Simian virus 40 (SV40): a possible human polyomavirus.
Vol. 95: Paris, 1997	Preclinal and clinical development of new vaccines.
Vol. 96: Washington, DC, 1995	Characterization of biotechnology pharmaceutical products
Vol. 97: London, 1997	Biological characterization and assay of cytokines and growth factors.
Vol. 98: London, 1997	Inactivated influenza vaccines prepared in cell culture.
Vol. 99: Strasbourg, 1998	Animal sera, animal sera derivatives and substitutes used in the manufacture of pharmaceuticals: Viral safety and regulatory aspects.
Vol. 100: Geneva, 1998	A celebration of 50 years of progress in biological standardization and control at WHO.
Vol. 101: London, 1998	Alternatives to animals in the development and control of biological products for human and veterinary use.

Developments in Biologicals

Vol. 102: San Francisco, CA, 1999	Advances in transfusion safety.
Vol. 103: Annecy, 1999	Physico-chemical procedures for the characterization of vaccines.
Vol. 104: Langen, 1999	Development and clinical progress of DNA vaccines.
Vol. 105: Paris, 2000	Progress in polio eradication: vaccine strategies for the end game.
Vol. 106: Rockville, MD, 1999	Evolving scientific and regulatory perspectives on cell substrates for vaccine development.
Vol. 107: London, 2000	The design and analysis of potency assays for biotechnology products.
Vol. 108: Langen, 2001	Advances in transfusion safety - 2001.
Vol. 109: Washington, DC, 2000	Biologics 2000 – Comparability of biotechnology products.
Vol. 110: Palm Cove, 2001	Orphan Vaccines – Bridging the Gap.
Vol. 111: Utrecht, 2001	Advancing Science and Elimination of the Use of Laboratory Animals for Development and Control of Vaccines and Hormones.
Vol. 112: Bethesda, MD, 2001	Immunogenicity of Therapeutic Biological Products.
Vol. 113: Berlin, 2001	Process Validation for Manufacturing of Biologics and Biotechnology Products.
Vol. 114: Ames, IA, 2002	Vaccines for OIE List A and Emerging Animal Diseases
Vol. 115: Bergen, 2002	Laboratory Correlates of Immunity to Influenza – A Reassessment
Vol. 116: Los Angeles, CA, 2003	Development of Therapeutic Cancer Vaccines
Vol. 117: Langen, 2003	Consideration of Alternative Licensing Procedures for Vaccines for Minor Species, Minor Indications and Autogenous/Autologous Products

IMPRIME
RIE MEDE
CINE m+h
HYGIENE

novembre–2004